NEPTUNE SPEAKS

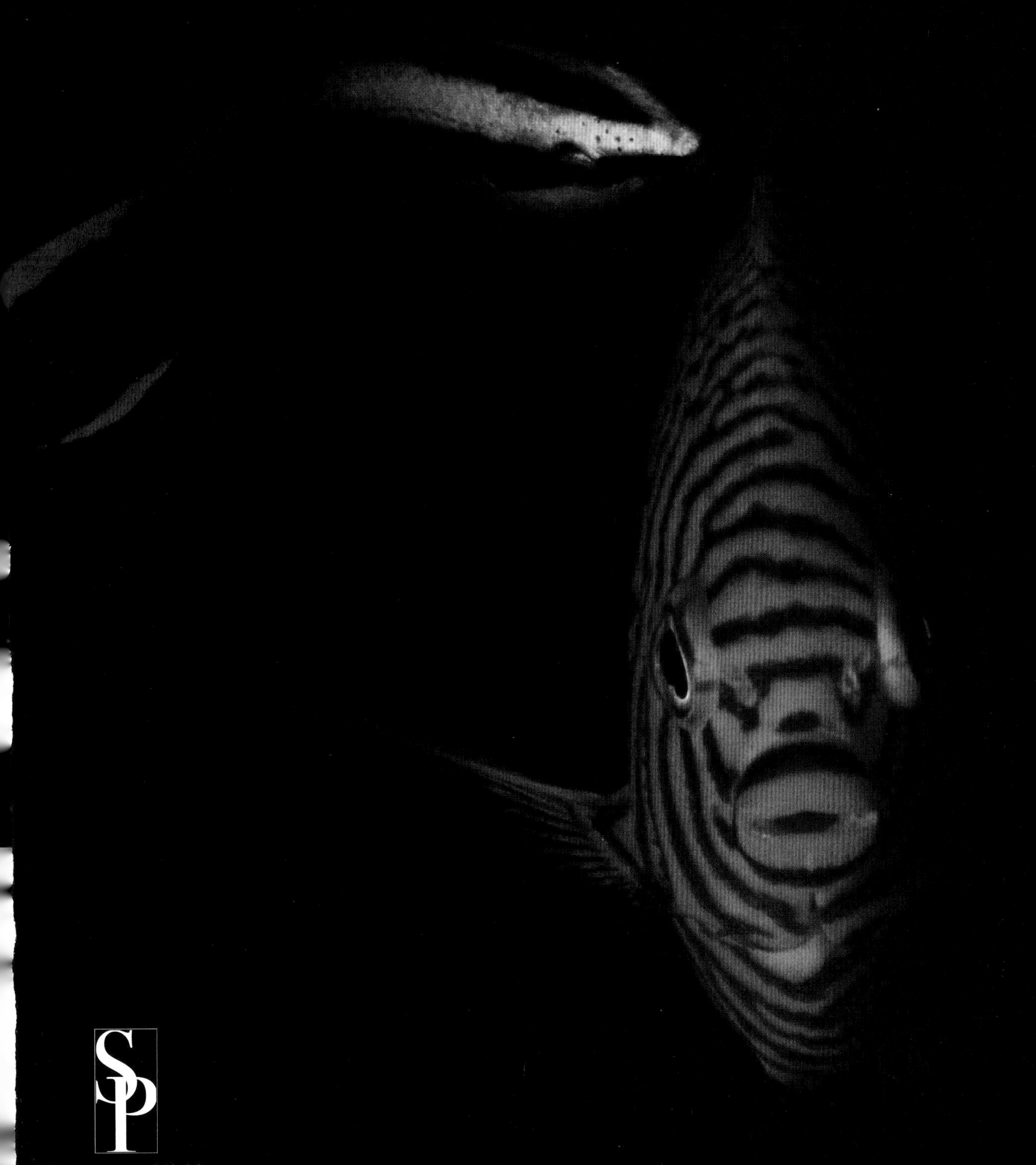

SP
Skyhorse Publishing

Skyhorse Publishing books may be purchased in bulk at special discounts for sales promotion, corporate gifts, fund-raising, or educational purposes. Special editions can also be created to specifications. For details, contact the Special Sales Department, Skyhorse Publishing, 307 West 36th Street, 11th Floor, New York, NY 10018 or info@skyhorsepublishing.com.

Skyhorse® and Skyhorse Publishing® are registered trademarks of Skyhorse Publishing, Inc.®, a Delaware corporation.

Visit our website at www.skyhorsepublishing.com

10 9 8 7 6 5 4 3 2 1

Library of Congress Cataloging-in-Publication Data is available on file.

ISBN: 978-1-62087-294-9

Printed in China

Acknowledgements:

Thanks to the waterdogs and gillbreathers of Maui, Tahiti, Fiji, Palau, the Great Reef, the Virgins and on around, whose minds and hearts share one reef love. Special thanks to Anita, Keith, Joan and Lulu for logistical coordination and support.

For whatever it takes in spirit, politics and getting the word out, thanks to George P, Peter R, Rene, Keiko, Inga, Teresa T, Wayne N, Kai N, Brenda F and all the troops digging in to save reef critters around the world, keeping wildlife in the wild, safe and free from glass enclosures and those who would put them there.

For the sun & the moon & my dog Lulu

ON WILDERNESS VALUES & SUSTAINABILITY

Neptune Speaks for wilderness values. The indomitable spirit is azure clear on pristine reefs, or it rages in heavy weather. Neptune loves fish in abundance but scoffs at "sustainability" and "best management practices." Who do they think they are, these self-proclaimed stewards of Neptune's clan who would put ocean communities on life support to ensure cash flow? Measured survival falls short of optimal balance. Wilderness abides by no extraction. Indigenous peoples may participate in the food chain, but commercial extraction is a compromise to wilderness, a risk calculated for profit. Reef survival becomes a by-catch, maybe, with the end of abundance a certainty.

They can't.

Examples sadly abound, yet we gain momentum in our daunting return to grace.

Neptune Speaks to the peril of trafficking in wildlife for the pet trade. Why are colorful reef fish still taken in Hawaii with no limit on the catch, no limit on the number of catchers and no constraint on rare, endemic or vanishing species? Because the trade makes big money—and spends it on "conservation" organizations who green-wash the trade. These outfits want your $upport to help with $u$tainability. You know them by their junk mail. Just as missionaries saved the heathen through resource allocation, some "non-profits" preach redemption through donation and sustainable trade. With directors directly vested in land development, banking, resource extraction and THE AQUARIUM TRADE, these ocean evangelicals blow green smoke up the collective anus. Moolah greenies beg the question: How can vested interests determine what's best for wilderness?

Lantern Toby, South Maui

Neptune Speaks through his minions of the deep, gill breathers sharing insight with those who will see, hear and feel what is going on. *Neptune Speaks* truth to data spun shamefully for personal gain. That is, truth expounded herein by a colorful chorus in a singular refrain. Harmony resides in wilderness. Wilderness is wild, with God-given abundance that no human person should put asunder.

Humanity, curb thyself.

Hawaii, stop the aquarium trade from stealing your reef wildlife.

Endemic species are found nowhere else. Relic species are endemics with no other species even similar. The blueline butterfly on page 4 is a Hawaii relic, extremely rare and vanishing. The Hawaii Department of Land & Natural Resources says the blueline butterfly is "experiencing a 100% decline" on major reef segments of the Kona Coast. DLNR calls them "a species of concern," because "we can't say why they're disappearing." The DLNR director is a former aquarium hunter. He resists a ban on the aquarium trade. The blueline butterfly is for sale on-line for $120. Shame on the State of Hawaii and those who plunder Hawaii reefs.

As long as the venal scourge persists like a rash, we will reach and scratch. Not every crime against nature is brought to justice, but few victims get more support from humans than reef wildlife do. Not all my friends are fish, though my dance card leans in that direction, which may be another story with more Rorschach analysis.

The octopus ink to the left reminds me of a mermaid's bomble. Does that make me unstable? Or worse: mean-spirited?

Things get murky on the psychiatric level. Some humans test lower than wildlife on compassion and a sense of balance. Aquarium collectors seem devoid of these fundamentals, rationalizing their vile extraction as necessary, convenient and harmless. If $u$tainability is your hue and cry, if you lie, spin data and oppress reefs for money, you a ho. You justify being a ho by denying truth that is self-evident to the rest of society.

I, Snorkel Bob'm unloved by the aquarium trade. They want to make me the issue, to distract from the indisputable fact of their reef-killing extraction. I am called the anti-Christ, and Snorkel Bob Round Pants—the president of Kona Aquarium Scourge called me, Snorkel Bob, an eco-terrorist just before Willie Kaupiko caught him reef raiding way inside the no-take zone at Milolii, the last working fishing village in Hawaii. Willie caught him AGAIN 3 weeks later. The issue is crime against nature. The aquarium trade is devoid of aloha. I too have cast a few stones, comparing aquarium traders to the hapless fellows on Predator Raw who hang their heads when informed that their wicked appetite for a little brown round is now being broadcast on national TV. I didn't actually call aquarium collectors and resellers pedophiles, but they took it that way.

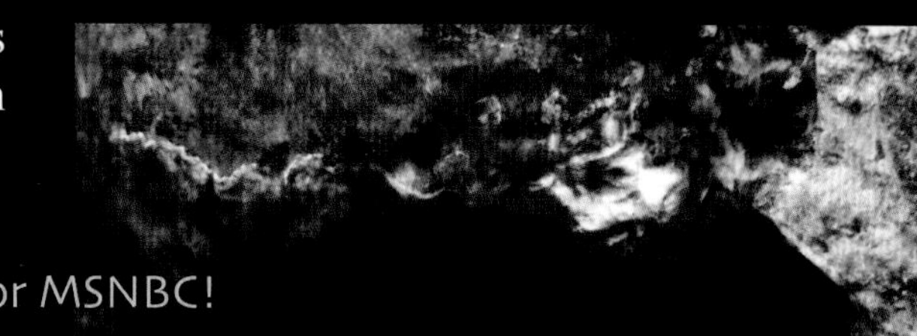

Hey. I'm not a gerbil. I'm a reporter for MSNBC!

A dwarf moray eel is 7" full grown. These colorful, spritely critters favor tiny burrows in current, because home delivery is the better part of valor with predators around.

I've seen aquarium collectors pull up to a dock with holding tanks full of wildlife for the pet trade. They shun observation, like the guys on Predator Raw—I'm just saying. One aquarium-trade reporter complained of no balanced dialogue on a ban in Hawaii. An aquarium magazine said I, Snorkel Bob'm to blame. Said I, Snorkel Bob: "Balanced dialogue? On one side you have 99.7% of the people of Hawaii. Let's call that side a very big, shaggy dog. On the other side you have a handful of collectors—we'll call them the fleas on the shaggy dog's ass. You're suggesting that a big shaggy dog engage the fleas on his ass to determine just how much longer and how much blood those fleas will be allowed to suck. The problem is: no self-respecting dog will engage the fleas on his ass in any kind of dialogue. That dog will scratch his ass to get rid of the fleas."

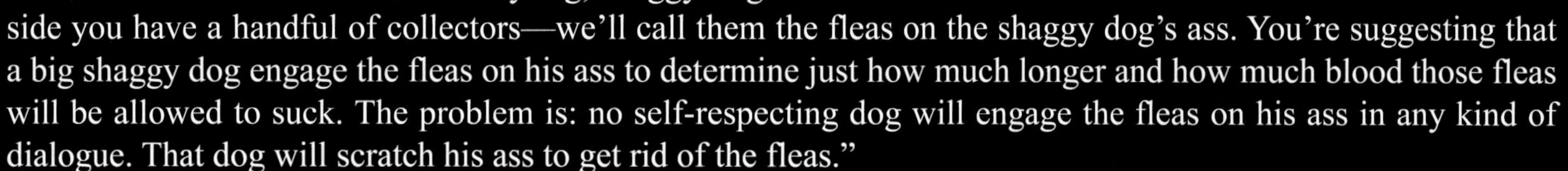

I thought the imagery vivid, providing a good laugh for all parties, with razor wit and rapier insight. But again, tsk tsk—they thought I'd called them blood-sucking parasites. Maybe they are sensitive. I hope so.

With apologies and contrition I have pledged to avoid the dog/flea illustration in the future with sincere hope that the aquarium collectors and resellers will jump off.

Flanneljammies reflective:

Boy, Snorkel Bob. You think I push the style envelope in my flannel pajamas, but YOU got *chutzpa*!

NOTE: though all blennies are forthright and unequivocal, the flanneljammies blenny is notoriously frank.

NEPTUNE SPEAKS

ON SHOW BIZ & REEF ROMANCE

I feel much better—it's best to clear up misunderstandings so we can all be friends again. It goes to show that Neptune's message is not all hearts and flowers. It can be harsh, like storms at sea. It can be sad, happy, despairing, joyful or as rich in pathos as a boy from the Midwest looking for a break on the mean streets of Hollywood. Shift that image to a few feet below sea level, and you have an eel from the mid-Pacific looking for a break on the sweet reefs of Maui.

Kukla came on with a streak o' razzmatazz, cruising up between my longtime snorkel buddy Matt Roving and me, Snorkel Bob, along the north bank of the sand channel at Molokini Crater. We didn't mind but suspected ulterior from the get-go. He was friendly, but you don't get a big hunky eel coming on like a BFF from way back, if you got no cameras. Kukla only made his move where he could. He's built for drama in the 7' x 1' range. He comes on casual, I'd say about 25 years old, judging from dental wear and a few gaps. But he cruised on in like a production assistant, like maybe we could grab lunch or something. Then…it happened!

Out of the blue, Kukla pitched a script. I couldn't see much there, but the eel believed. He called it a tale of reef intrigue. We had ½ hour or so, and he did have us pigeon-holed, so we shot it:

Left, dragon eel; below, Kukla cruising

Volume 1, No. 1, Kukla Comix, starring...

Kukla the Yellowmargin Moray

& co-starring Cleanerboy*

*nominated for best Hawaiian cleaner wrasse in a supporting role

FADE IN: Any day on any reef, with critters who need cleaning. Nobody does it better than Cleanerboy, and nobody loves the attention more than Kukla.

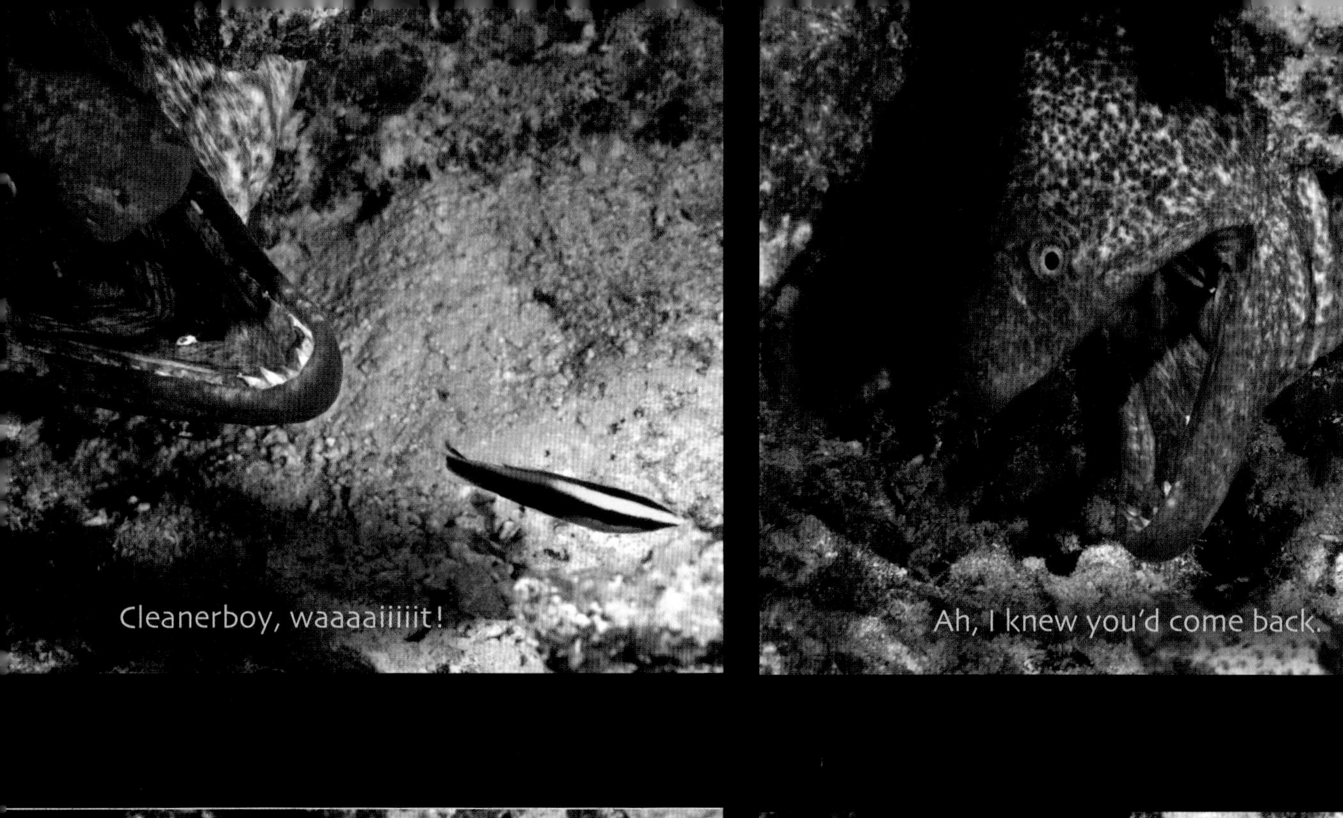
Cleanerboy, waaaaaiiiiit!
Ah, I knew you'd come back.

Aggghh... Who?

Cleanerboy? No, I haven't seen him. Why?

Like I say, pathos—with ethos and chaos, and we wish somebody would pay us. Kukla Comix may go nowhere, but it doesn't matter. Kukla is squirming for joy, which can be a push with such a hunky fellow. Anyway, he's working on his next script. Frankly, big friends can be comforting on a reef—to a point. Now Kukla wants to direct. He thinks the world is ready for eel drama, and Clark Dragon will be perfect for his epic: *Jonathon Livingston Sea Eel. Love means never having to say I'll eat you.* Talk about a push, and I must say, Clark Dragon is not nearly as big in person:

I first saw a dragon eel in a Honolulu wharehouse (yes, I can spell warehouse, this was a wharehouse) darting side to side, awaiting shipment and/or death. Dragons are sold on-line for hundreds of dollars “to those aquarists who like eels.” Compassionate people who like eels will be snit outa luck if the aquarium scourge has its way. We’re keeping it mum on whereabouts for dragons and others. If rare wildlife is taken for the pet trade, that wildlife will disappear. I will not compare aquarium collectors to Gestapo agents here, though it is sorely tempting.

Wait a minute—with Clark Dragon in the lead, who'll play the love interest?

Meeeeeee!!!!!

Cut! Cut! Cut! When a blue ribbon eel is blue, he's a male. Females are yellow with opposite trim. So how can he be the love interest? Well, we could, er, uh, make this uh…

Yes! *Brokeback Reef.* We'll be nominated! We'll go to Fiji where she lives—I mean he—because extras are so plentiful there, like these garden eels. I, Snorkel Bob've approached a garden eel or 2. They duck into burrows. But Fiji gardeners aren't shy.

Speckled garden eels, Somosomo Straits, Fiji

Wait! I got it: *My Big Fat Eel Wedding!* Cozzin Freckles got jilted and still twitches over lost love. He's an original sourpuss who couldn't help but get that smirk off his face. He may be Royalty, which makes him the perfect fall guy in a romantic musical dramatizing the endless circle of reef love…

Freckles the freckled snake eel catching a few rays shallow. Taciturn to a fault, at times morose, Freckles is known to loosen up in certain specific company...

…because this is showbiz and could be BIG. Freckles nearly palpitates over a little box who has no thought for anything but snax.

Speckled Beauty? No. That seems forced. She was cute but hardly beautiful.

Hawaiian Boxfish, aka Boxgirl, assessing snax potential on the ceiling.

Meanwhile, it's a shoo-in on the plot, with Boxgirl preoccupied on upward snax.

Boxgirl, working the upper schmutz.

The fall-away jumper.

Boxgirl in the groove.

Friends and family worried that she'd plump up like a sea slug, but as she bulked up she got pretty in a cubish way.

Meanwhile, hardly a drop and a dribble below, Freckles watched forlornly, a love-struck freckled snake eel with low odds on romance requited.

Freckled snake eel youth

But Boxgirl noticed Freckles as she filled out. She found him curious, possibly drawn to his…freckles….

Tributaries most often join the greater flow. In hermaphroditic transmutation the dominant female of some species will come out, as it were, when the male departs, as males of many species will do. Boxgirl, a Hawaiian boxfish, went yellow where her darling dimples used to be. The voice dropped as the nose bulged with whiskey protuberance.

Hawaiian boxfish male

Before you could say *Swimming with the Stars*, it was purple flanks, belching out loud, an attitude and you couldn't tell him a thing...

Et voila...

Box Macho

From there loose ends tied up easy as spaghetti straps on an evening gown. Freckles hooked up with Box Macho in the first ever BLT community transgender, true love epic musical comedy/drama with a happily ever after for all.

I am humbled. I want to thank the Academy, my parents and teachers, all my cozzins and in-laws and all the little fish who gave so much to make this possible. I'm not sure what comes next. I might do a stint on 60 Minutes. I'm only 80 in eel years.

Boy, showbiz. Who doesn't get carried away? One little comic book and we're off to 2 epic movies. Hey! How did you get past security?

Pardon me! But with an epic wedding scene we'll need a rabbi. That was me in *Bye Bye Braverman*. Or a lawyer for the pre-nupt scene. I can do that! I was understudy in *The Eels of Eastwick*. I am not your average whitemouth moray!

Look at me!...

Marlene is a yellowfin surgeonfish at Enenue, the Beverly Hills of Molokini Crater near Maui. She won't say what year she came on the scene, but she goes way back. She'll drop anything for a screen test. How long can she pose? How much time you got?

You have a nice touch
for such a young wrasse.
You must know how I feel about you

ON HUMILITY

Sorry. We will find a part for Marlene and review the eel's resumé. That's the thing about greatness: everybody wants some. But if you get some, it's addictive, demanding higher dosage at greater frequency. Take it from me, Snorkel Bob, humility is daunting but necessary in a world begging for guidance. Voices in the wilderness must unite in a call for wilderness values.

What is wilderness? If an aquarium collector bags a few yellow tangs, and nobody is watching, were the fish ever really there?

From a Molokini sunrise to noon at Ukumehame to dusk on the Westside, these yellow tangs are making a comeback, keeping Maui reefs alive.

What's the difference between wilderness and a void? Wilderness is rich and rife with wildlife, while a void has the dull vibration of "growth." Waning wildlife is a first symptom of a void, along with aggregating sheetrock, shag carpet, drop ceilings, asphalt, convenience to church, schools and shopping and of course much, much more.

Young parrotfish (above) and kala or unicornfish (right) cruise as adolescents often do. Maui alone in Hawaii banned gill nets a few years ago. Now we see parrotfish and unicorns schooling near shore, giving ever more reason to do it for the children.

A tiny parrot at ¾"... *a teency shortnose wrasse at ½"...* *& a teency tiny shortnose at ¼".*

Juvies frolic on Maui reefs that were void a few years ago, because more fish spawn since the gill net ban. Troglodytes with scuba gear or surface-pumped hookas still spear north shore parrotfish at night, voiding entire reefs for a special sale. It's called "fishery management" at the Hawaii Department of Land & Natural Resources.

Where was I, Snorkel Bob? Ah: humility's daunting challenge with so much to disturb us—but we *hele* on (heh·leh, to proceed) with miles to go before we sleep.

Sage advice is rare. Kukla still looks young in his promo shot (right), with the vigor, innocence and perfect choppers.

Come on, kid. Try to relax. We're in this together.

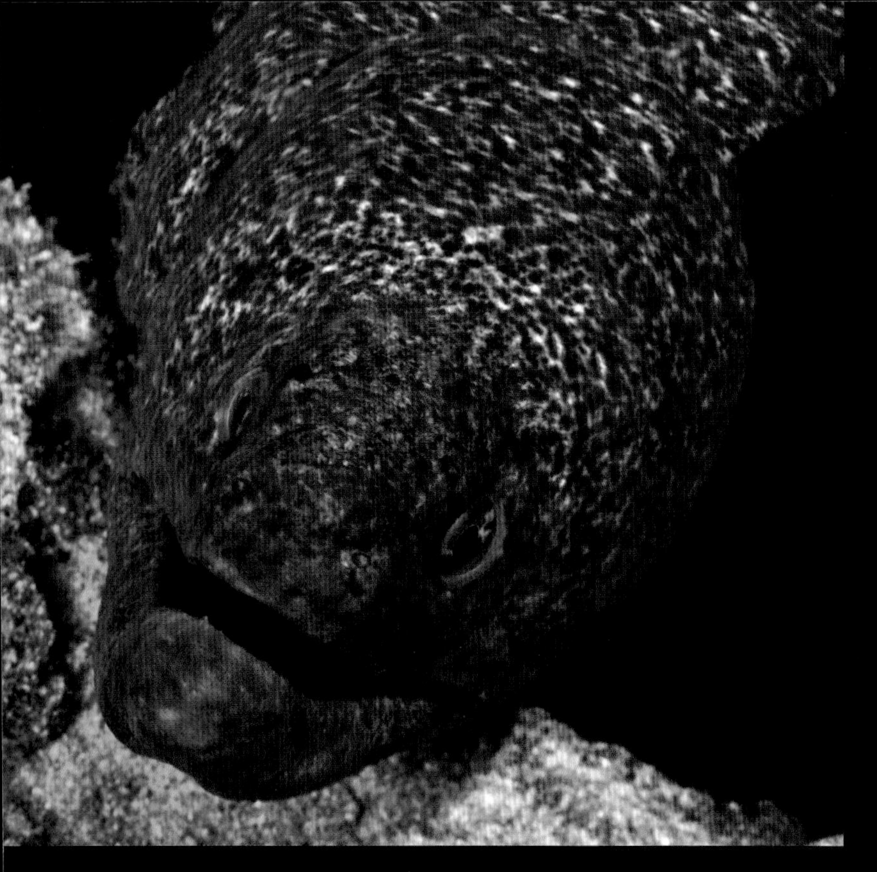

This is Kukla today, telling me to calm down and have faith. He's a little long in the tooth—make that missing a few teeth. But we've been cruising together for years, and it hardly gets better than an old eel who senses needs even before you do.

My friend Ziggy is a world-class waterdog in transit from youthful vigor to sage wisdom. He advised before a dive in Cane Bay, St. Croix, "I'm going over." The Cane Bay bottom descends to 95', where the ledge drops sheer to a few hundred. "Not too far. One thirty-five. One forty maybe. Okay?"

Why?

He shrugged. "I need to feel it." I declined; no need. I'd felt it. It's pressure—nearly 5 atmospheres. Worse yet, a tank goes quickly at that depth. We affirmed at 95', and over he went as a bull shark cruised the yonder murk, a big beast about 13' long and 4' thick who easily dismissed a couple of snack-size white pecks.

What did Ziggy feel?

He shrugged again. I suspected he felt the pressure and hazard of 140'. I suspected that he'd "needed" these unique feelings to match his unique status as a waterman.

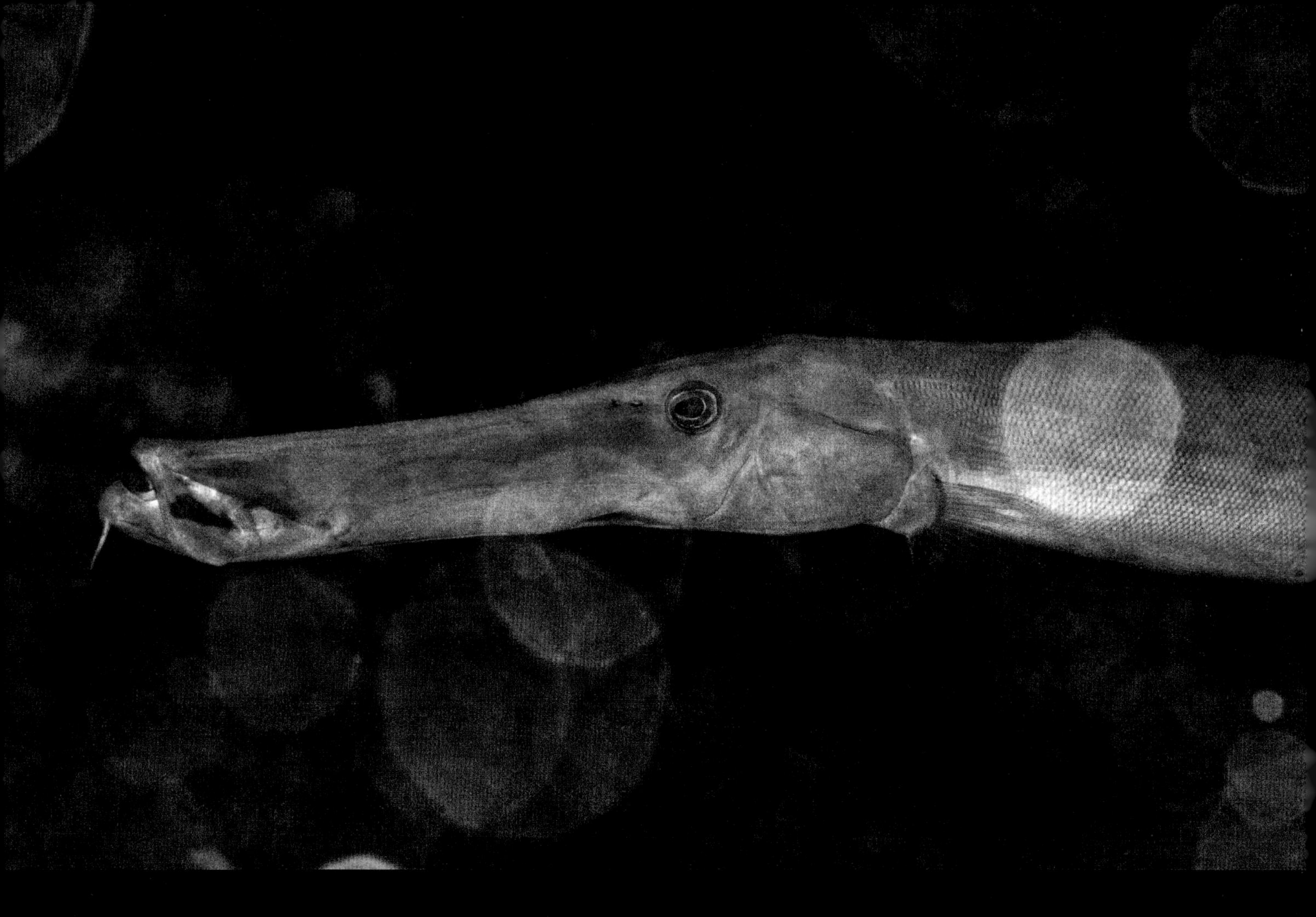

Nitrogen narcosis or rapture of the deep can begin at 70' with euphoria and insight to the meaning of life, with no need for mask or regulator. With blissful clarity, breathing between the water, down go the rapturous.

This trumpetfish is in a dark, silty cave. It looks more like acid flashback than narcosis, but any mystical overlay at depth may call for realignment.

On the other hand, a raccoon butterflyfish checking your pupils for excess dilation is neither obtrusive nor a concern but an example of a dive buddy doing her job.

Raccoon confidential

Oh, Neptune! What happened?

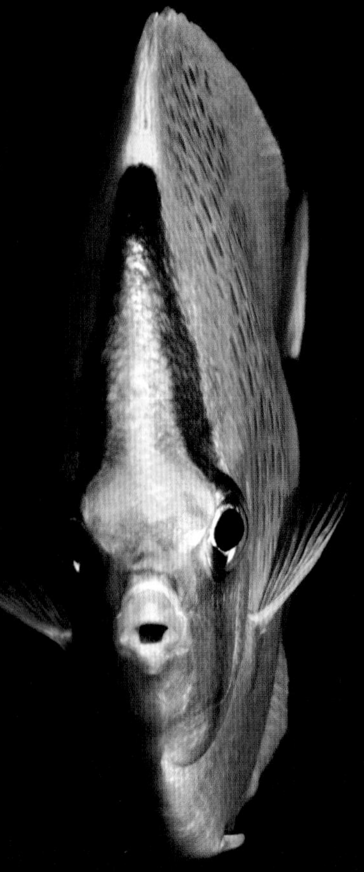

About a year later Ziggy told me of his next peril at 150' in open ocean—no wall, no bottom. The group wanted predator footage in the water column, big billfish like marlin, sailfish and swordfish and maybe a shark or 2. Thermoclines are temperature gradients between layers of water. At 130', below several thermoclines, the group disintegrated in a deep, swift current requiring calmness. Down-drafted to 140', Ziggy focused on even breathing and the upwelling of his salvation. It did not come, nor did it come, and then it did not come. He felt no difference between 140' and 150', except for fading calmness, colder water and growing fear of 160' or 180' or 3,000'.

The upwelling never came, but like the soft touch of God, the current simply ceased, allowing a waterman of significant instinct to rise slowly as his slowest bubbles. Even as his gauge showed inadequate air to make the surface, he knew the air in his tank would expand on ascent to allow more time. Rising from 150' requires a decompression stop so nitrogen can off-gas more slowly to avoid the bends, in which nitrogen jams the joints and contorts the body. Ziggy hung out at 20' for two minutes till his tank bumped up next to empty, then rose again with faith or desperate hope that he'd ditched enough nitrogen from his blood stream.

With calmness falling just shy of adrenaline, he surfaced to open seas with no person or vessel in sight.

Don't stop now! Then what happened?

Well, that's the end of the story. Ziggy got picked up in a while—the boat had to fetch those in greater need farther away. He didn't get bent, though he avoided hot showers, carbonated beverage or stimulation of any kind for a day and a half.

Where did you go wrong?

"I didn't like it, man. I didn't like it. I didn't like it. I didn't like it. I followed a bunch of idiots. They didn't know. They said they knew."

Catharsis seldom needs pressing—but try to tell Goatboy about nuance...

Ziggy thought it over and sighed, "I don't know, man. I don't know."

I didn't chide my friend that clutch instincts can be wasted on foolish company, because experience had yielded an upwelling at last.

Amethyst anthius

ON WHAT A FISH MIGHT CONVEY

A bottom at 50-55' with moderate halimeda (oatmeal grass), will host a few regulars like the 2-spot wrasse, a subdued character in movement and coloration. 2-spot adults range 2-3" and blend with halimeda leaves ¼-½" in diameter—many greens mottle among many beiges, grays and browns in a still-life to fool a predator or photographer.

Hiding under the covers a young 2-spot wrasse peaks through to see what monsters lurk.

But a 2-spot can be distracted. Cavorting with exotics, exposed as a politician with a secret appetite, the adult 2-spot will freeze full flair as the big log drifts near and aims its camera. Though still demure in beige, light beige and medium beige with beige overtone and trim, the 2-spot lights up when strobes flash.

A 2-spot caught out (left) cavorting with a smalltail wrasse who knows how to work it.

Wait a minute—is she coming on to him? Or me? It might seem odd at first blush to imagine a cold & scaly warming up to a large & clunky, but check the body verve, the eye contact and oh, that yellow lip-gloss.

The smalltail wrasse femme is singular in vibrant charm and that certain j'ne sais quois.

You will NOTE her distinctive joie de vivre beyond the seductive flirtation; she is sincerely happy to see you, as if the money doesn't matter.

While the smalltail wrasse femme is irrepressibly coy, the male of the species is most often called a pencil wrasse. Also flamboyant—shunning wingtips and cardigans every time—he also seems resigned to her free spirit, perhaps confident of his critical role in her spawning needs. Though he differs in coloration from his femme, he matches her charm with aplomb.

Mr. & Ms. Pencil/Smalltail together, as sooner or later they must be.

So 2-spot wrasse gets caught out again, dateless and exposed, with no option but another full flair to discourage the hulk with the strange eye stalks.

Left, a 2-spot wrasse adult, 3" in full flair—dorsal, caudal, anal & pectoral—a stance in stillness may intimidate a much bigger photographer.

Full flair is a compelling behavior I practice in the mirror, just in case. 2-spot has another trick—flicking to a new spot. Then he flicks again to the halimeda brush. I followed him in and got him perfectly framed till he flick, flick, flicked…gone.

Humbled again at 55', nosed into a halimeda copse *con nada mucho* to show for it, the bottom verily asks: where is your wrasse now, Mr. Flair? Yet on a chance, mystical as Neptune's sibilant whisper, framed and focused in sudden perfection is a plea and a prayer; it's a sunbeam with a fish inside effusing gratitude, love and hope.

Well, you might say: *He hopes you won't eat him, and he might love you if you don't, but he is a frikkin' fish. And you're talking gratitude?*

Yes. I, Snorkel Bob, call them as they appear. A small green soul taking a turn in the halimeda will love her halimeda home and blended greens with gold filigree. She will await social calls from those who would protect the little green world from extraction, and yes, I think gratitude is Neptune's intention.

Maybe it's just me—along with minions of millions, gill breathers and lung breathers bonded in gratitude, love and hope…

The little green fish lives near a South Maui reef. Into a slight current across the halimeda flats to the pinnacle yonder, habitat is varied as Neptune's spawn. And so next we meet...

The Hawaiian seahorse male is green, the femme golden. She drifts through life on prevailing currents, hanging out here or there in gentle interludes. Halimeda is a favorite hitching post. This adult is 6-7" and may grow to 10".

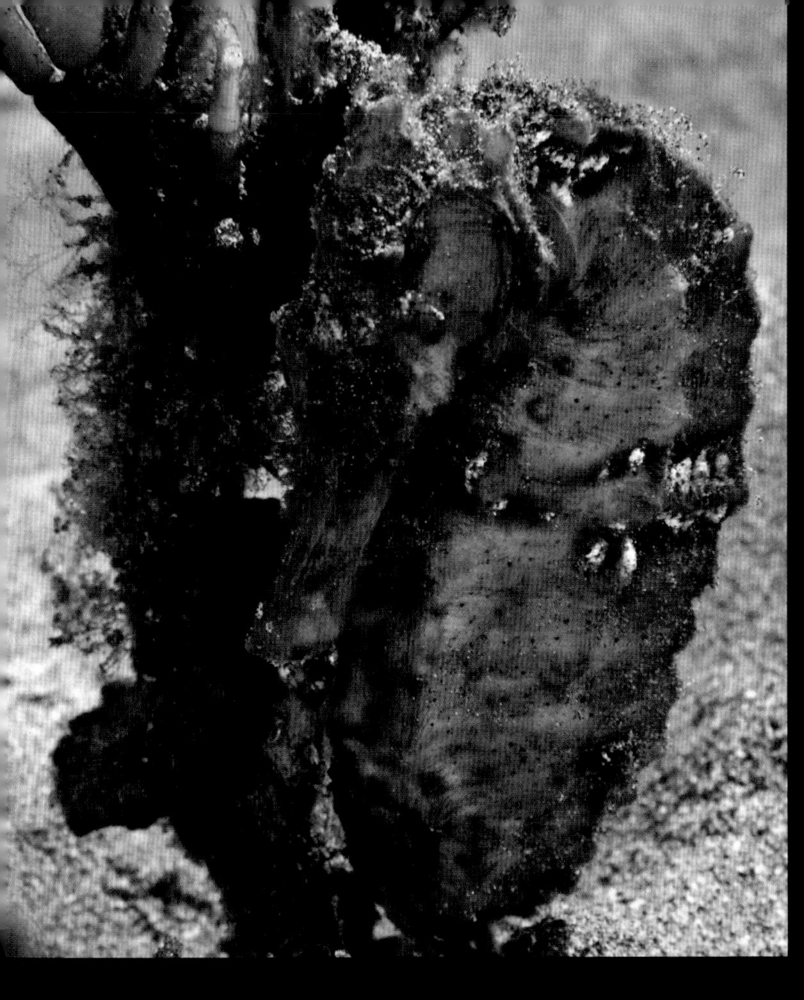

How many seahorses have I seen? So far in Hawaii it's 1 (one) per lifetime, sudden as a lottery winner. Can we compare cash windfall with spiritual communion? A seahorse fulfills a day and a life. Seahorses keep the secrets of the deep. As I rounded, she followed, also mesmerized by the creature out front.

The seahorse message is obvious if you see it and defies translation if you don't. A friend summarizes the message for lateral line comprehension. *Humuhumu* is Hawaiian for triggerfish. The best-known Hawaiian triggerfish is the humuhumunukunukuapua'a:

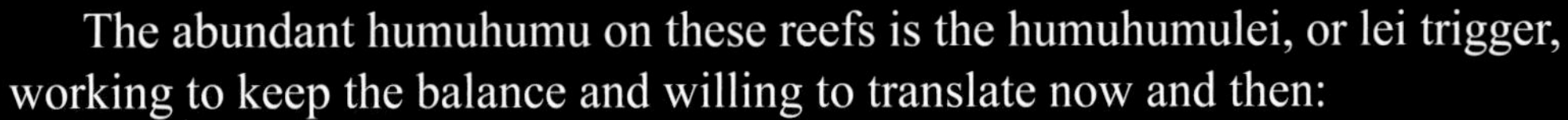

The abundant humuhumu on these reefs is the humuhumulei, or lei trigger, working to keep the balance and willing to translate now and then:

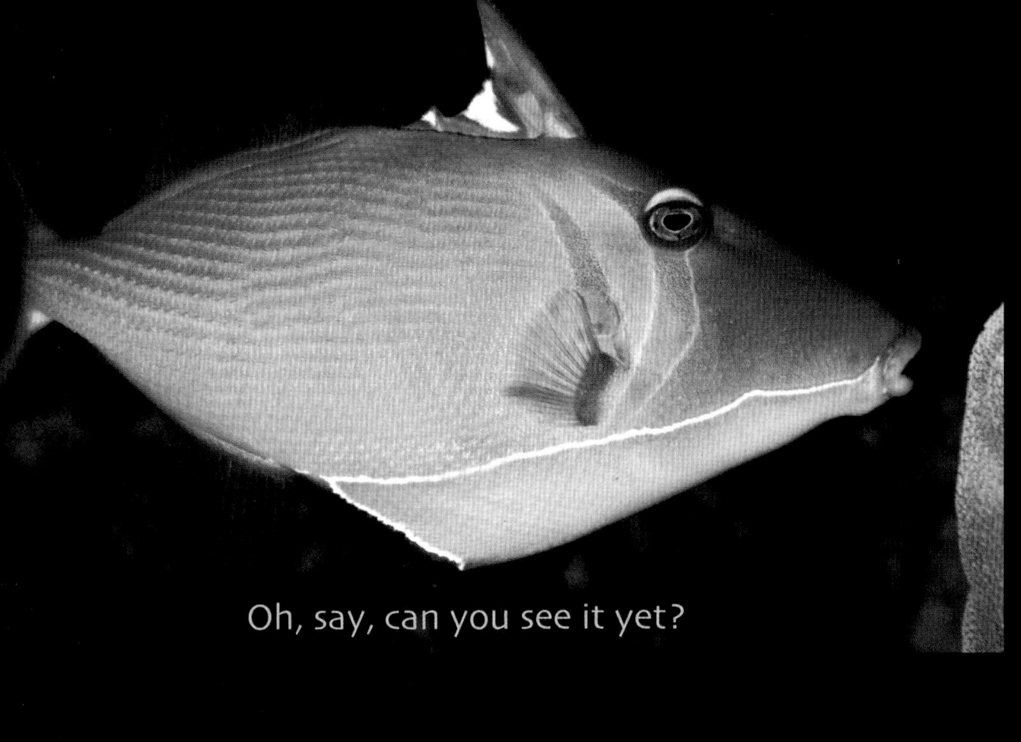

Oh, say, can you see it yet?

How about now?

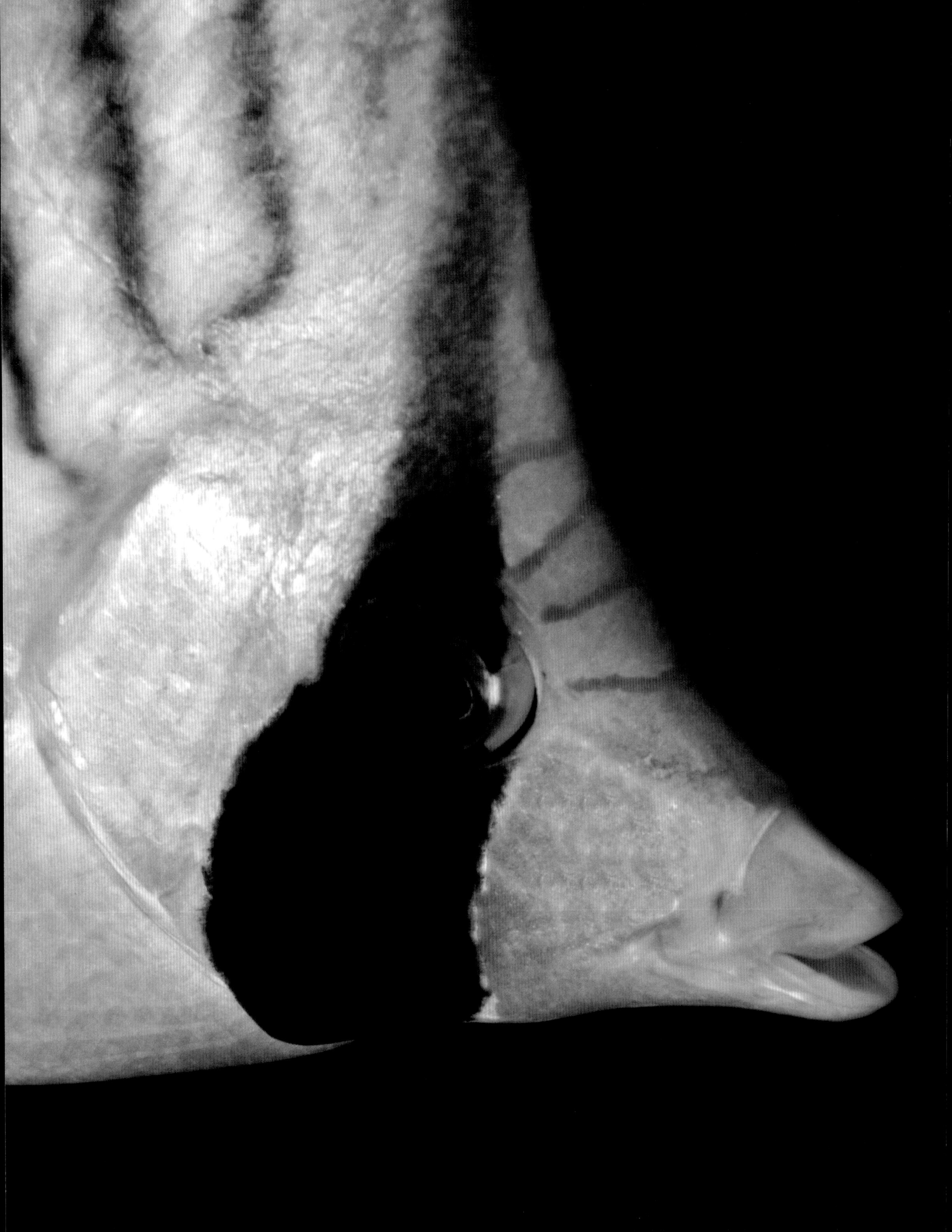

ON NEPTUNE'S WAY

Adrift, billions of tiny specs feed on ever-tinier specs while serving to feed slightly larger specs. Life's transition is seamless, giving over and into abundance.

Enter man—and women and children and convenience to church, schools and shopping. Neptune absorbed the topside overlay for aeons, till the human species spawned needs uncontainable as aberrant cells, needs to consume everything.

Wild critters live and die daily with the same fundamental drives—propagation, hunger and fear—that all critters share. The difference between wild beings and the organized species is that life's transitions occur with more profound resignation, with far less drama, ceremony or desecration to those who don't suffer a veil of tears but live joyfully one day to the next in sunlight and soothing rain.

Are wild critters anarchists? Though hard to organize, they achieve order and balance as yet unmatched by civilization. The operative word here is wild, a condition that is not forbearing on the cordial behaviors, like social, communal or cooperative.

Neptune's way is over and around, up and down and therein constant. A threadfin butterflyfish (left) enjoys its daily pursuit, keeping a reef balanced on South Maui.

Death gets many people squeamish. But fish live and die every day without the fuss or regret required by humans. We all join the inexorable current downstream. Wilderness people don't destroy what surrounds them or take more than they need or capture prey for amusement, while human behavior remains on the hole excruciating. I mean whole.

Where was I, Snorkel Bob? On Neptune's way, heading into a slight current to check out the yonder pinnacle, arriving at a small coral head maybe 10 or 15' long by 4 or 5' high with a *puka* front and center. (poo·ka, Hawaiian for hole. Remember puka shells strung as necklaces?) So we ease into the lee of the current, just as those who live here have done in choosing their home. What ho?

Yellowstripe coris: endemic to Hawaii means that it's found nowhere else on the Blue Planet.

It's an endemic yellowstripe coris, a young'un calling out, Hi, Bob! Or maybe she's only poised for another gob o' reef schmutz. Or maybe it's a bit o' both. Anyway, here's a paradox or 2: for starters, her stripes aren't yellow and will not turn yellow, even into adulthood. For finishers, she's endemic and extremely rare. She generates far more excitement than your common yellowtail coris (left), whose tail *is* yellow and whose personality may be the most exotically irrepressible, engaging and playful in the neighborhood.

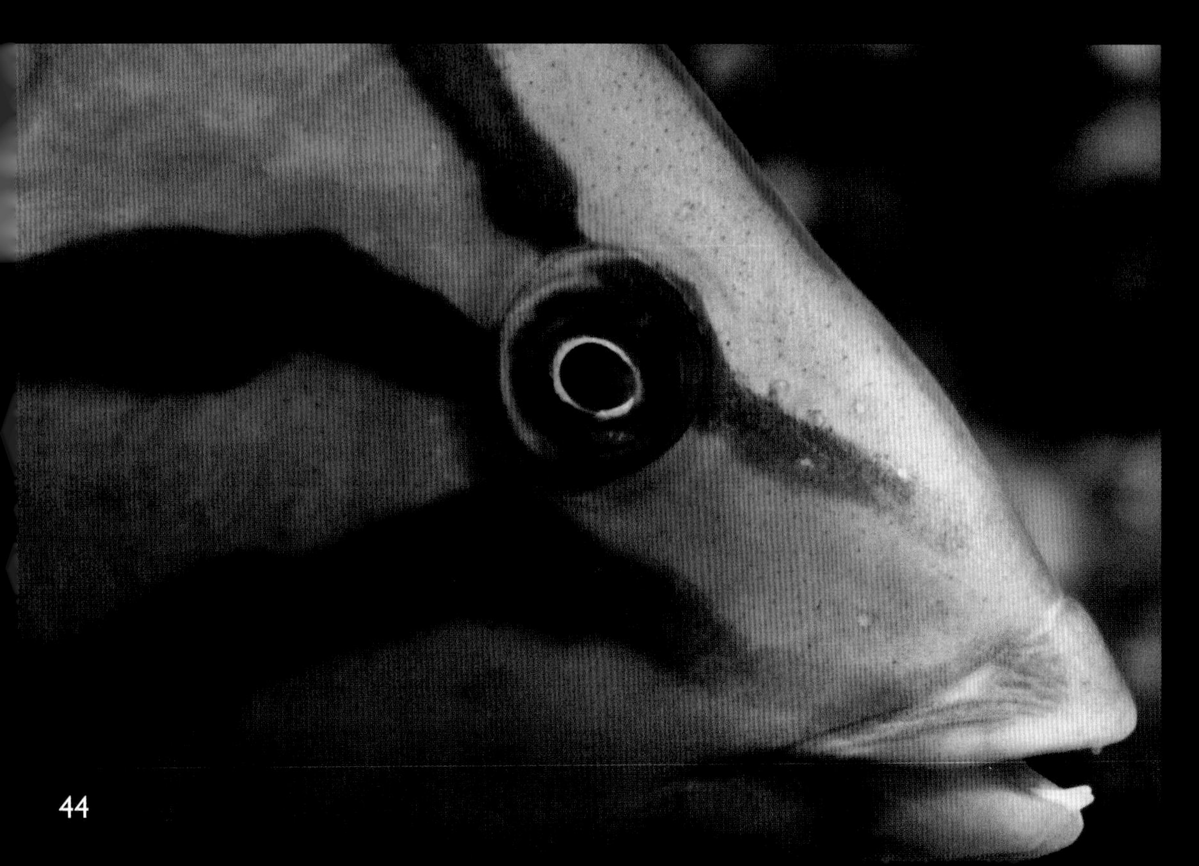

That was a yellowtail coris doe; here is the buck:

The young of most species are most endearing, and the youthful yellowstripe coris indicates trust by feeding in comfort.

Here's what the little tyke will grow into, Neptune willing. This adult female yellowstripe coris is part of a pair who live on Maui's north shore, all the way around from this juvie's home. Every area has different color qualities, from a greenish tint near shore in the south to photo perfect blue up north:

Yellowstripe coris adult, bottom and right

Back south in the green tint, moving into the puka we find an extended family making a life together. Out front is another youth, this one a bi-color anthius.

Anthius are rare now on South Maui reefs, which is better than gone, what the aquarium scourge made them until Maui County encouraged the scourge to be gone.

A bi-color anthius adult near the home puka

The Kona coast is photo perfect blue water and 135 miles of continuous reef cover.

Back on Maui's south shore we can see that a buck cruising solo…

…and a doe taking note…

May well lead to…

…anthius love.

Near the base of the coral head, a zebra eel watches the world swim by,

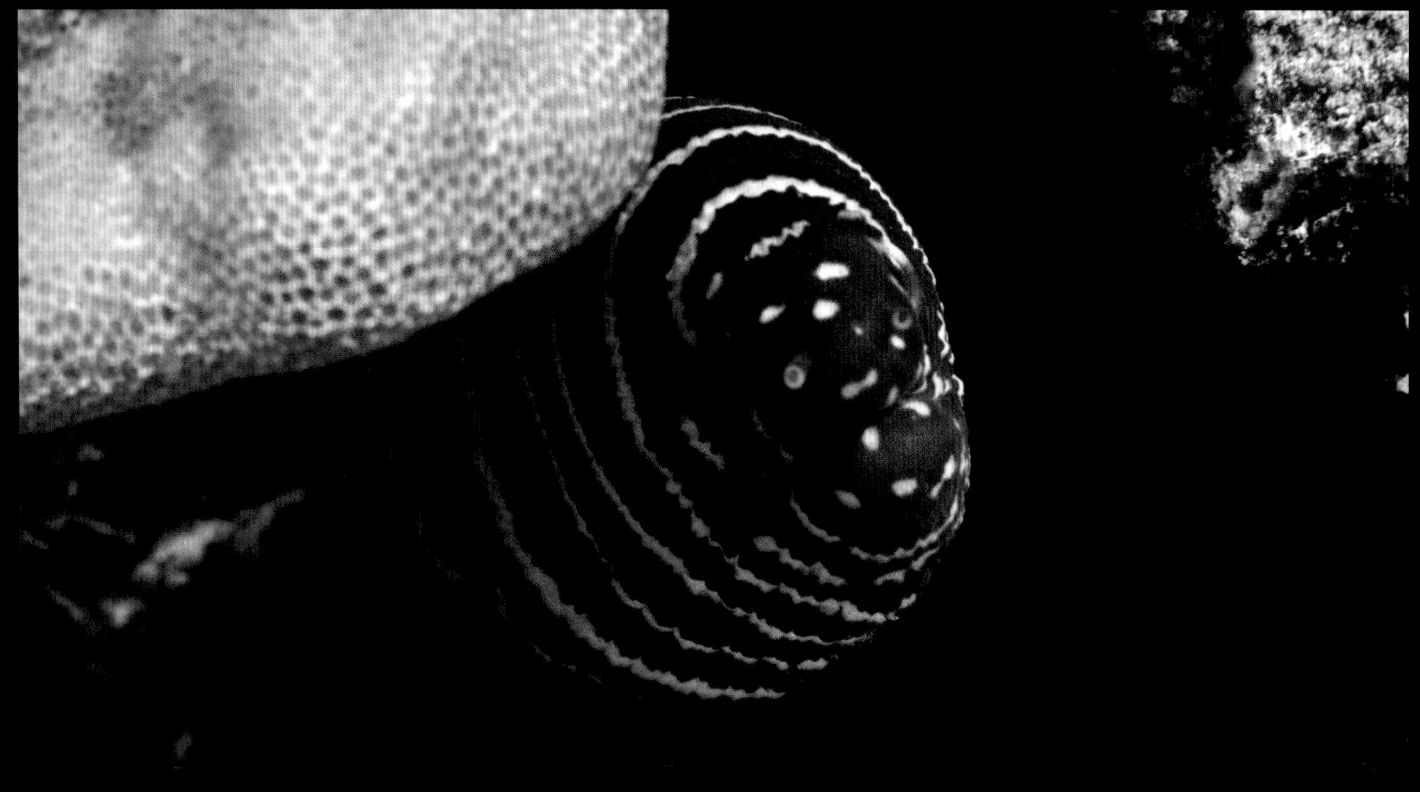

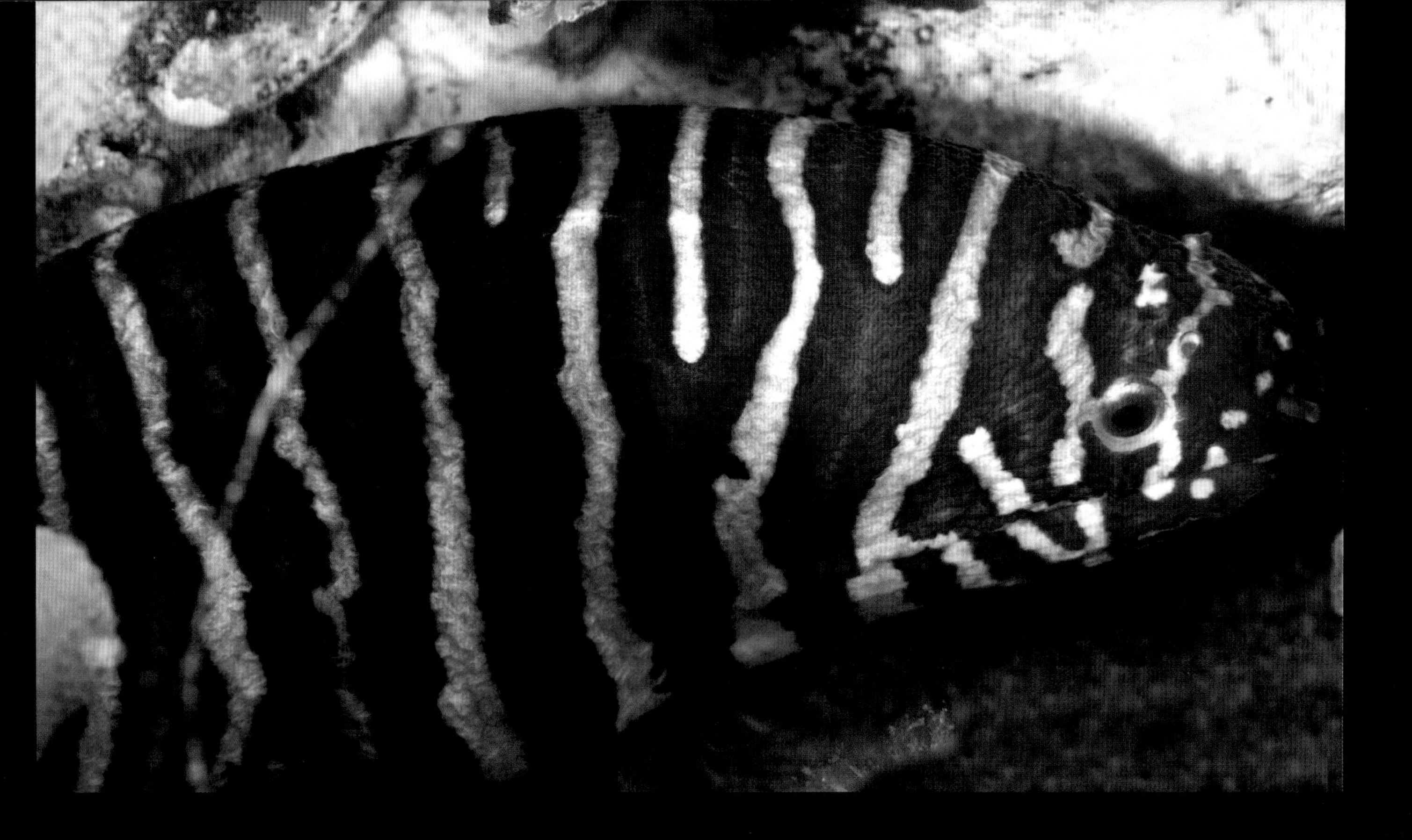

Or zebra could be seeking an appointment with the little emperor…

I am the Emperor of Swimps! I mean Shrimps!

A cleaner shrimp's self-esteem may be inflated, but the critters do line up in abeyance for liberation from parasites and goobers. NOTE here that the swimp with the red spinal stripe is a true cleaner, shown here reflecting on his empire.

I meant shrimp—and I meant longitudinal stripe, since some mollusks, crustaceans and politicos have no spine. In both reef and political venues, compensation is attempted with color and flourish. Unfortunately, the banded coral reef shrimp also dabbles in demagoguery. He too will clean occasionally to earn a hall pass from toothy predators.

Banded coral shrimp

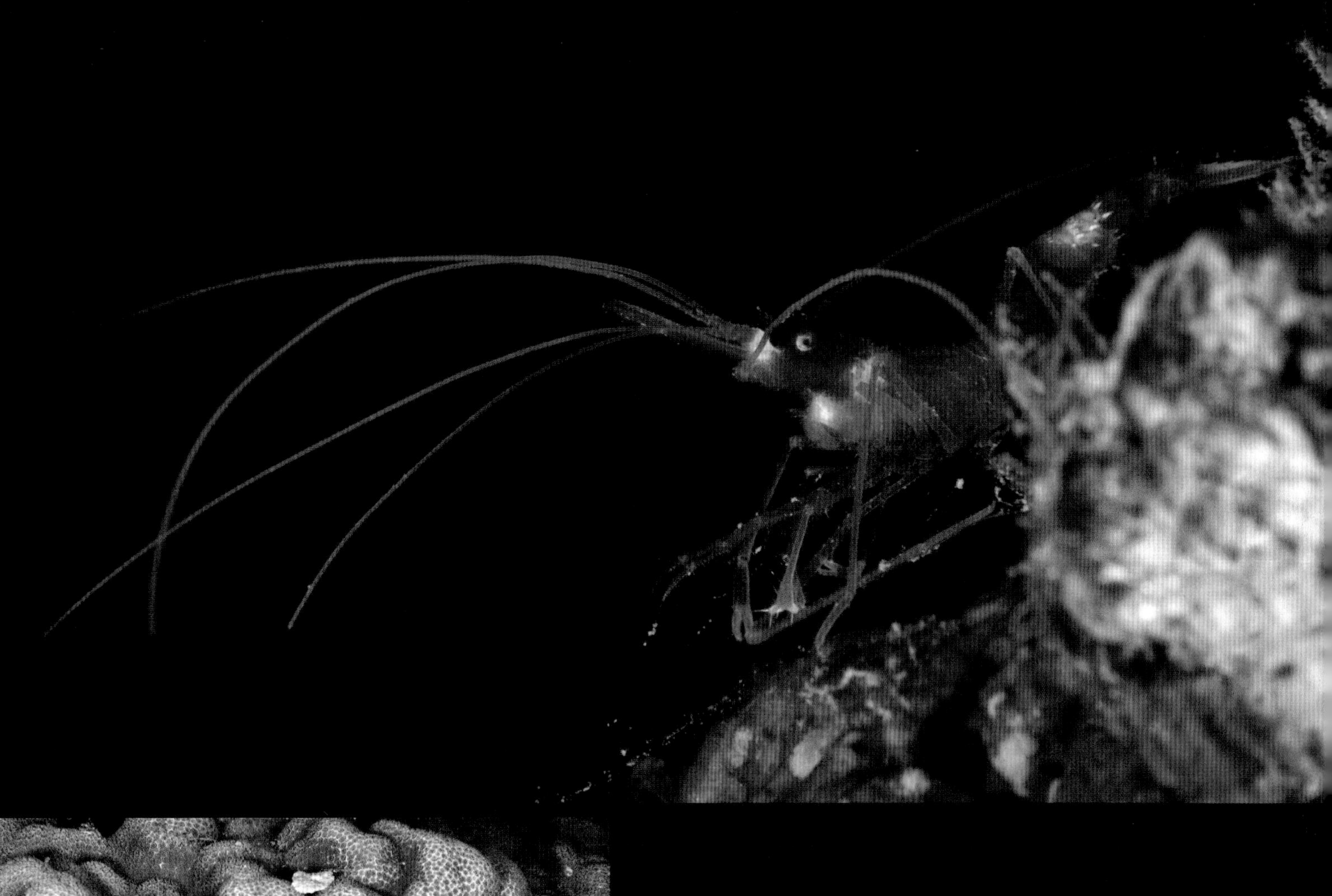

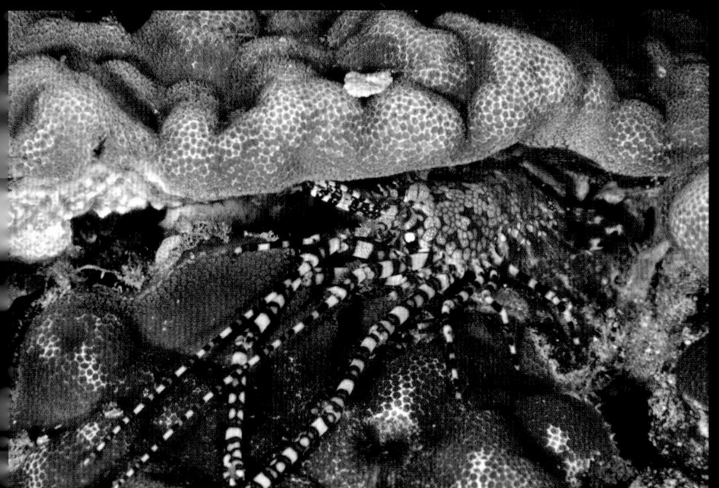

Bigger and more exotic is the marbled shrimp (left).

Spiny lobsters share optimal habitat to avoid predators and current—and to chew the fat, tell crab jokes and relax with a little seaweed.

Hey, I knew this crab—you couldn't get near her on account of the barbs on her claws…

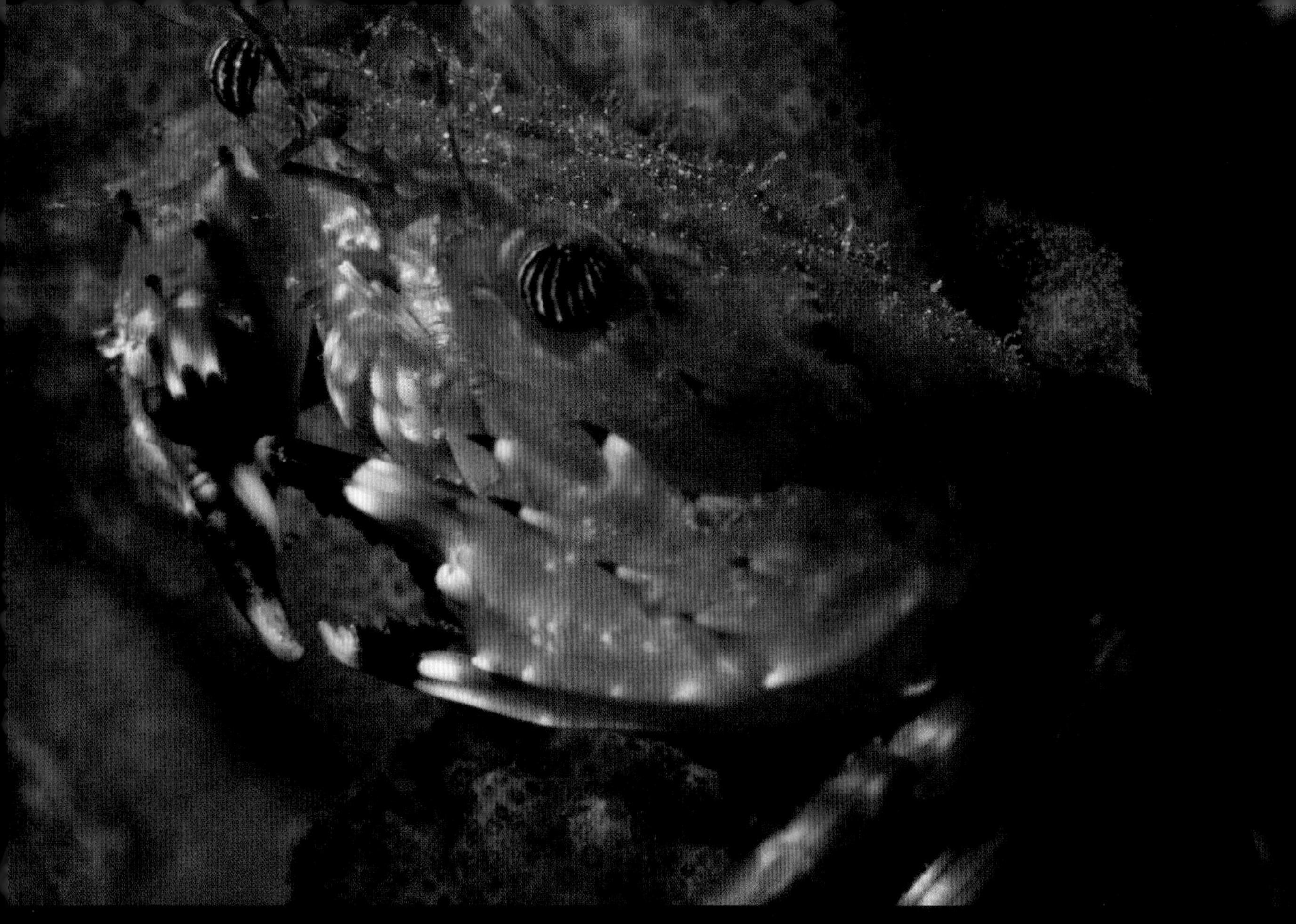

Some crustaceans stay grumpy, appearing on so many menus and in so many jokes.

Rounding the small but prolific coral head we meet an invert who causes alarm.

Crown of thorns starfish, close

Why does the crown of thorns starfish remind me, Snorkel Bob, of wife 1? Her skin was creamy smooth, so it must be the deadly neuro-toxins. Of greater concern than hazard to humans, however, is hazard to reefs. For aeons, these coral ravagers remained at 40' and deeper but now come in shallow to eat every coral polyp in their path. Climate change? Ocean acidification? Injection-well overload? All of the above?

Blue dragon nudibranch

On a happier, more graceful note, a blue dragon nudibranch wends her way up a coral-encrusted rock, sorting snax and detritus.

We come to a brief sandy patch with no coral, rock or halimeda, but you can bet your bottom dollar…

Bottom dollar

…that a snaxize cruiser minding her own beeswax can discover sandtraps (!) anytime. What can she do? That little pop you hear underwater is the sound of tiny fish sphincters slamming shut in the sloooow turn around.

Sandtraps. Reef cruising can be dicey as the 19th hole, with lizardfish and flounders nestled in the sand, ready to gobble now and ask questions later.

Often feared as reef gourmands, lizardfish are also prone to romance. Arriving at the near edge of the yonder pinnacle, it might as well be spring.

In the spring, young lizardfish fancy turns to love. Apparently compatible, this matched pair of lizardfish works out the pre-nupts… but…what ho?

We interrupt this cruise at the southern verge of the yonder pinnacle in the midst of invertebrate/predatory erudition for a suddenly thickening flow of algae or coral bloom or macerated sewage from a cruise ship—arrgh: these bits are squirming…

Better suited for a peephole than the big screen, a sea hare is a tiny nudibranch who drifts with many, many others.

Where was I, Snorkel Bob? Ah, cruising through the sea hares at the edge of the pinnacle, in rocks well-suited to an ambush curmudgeon.

You could throw something in the hopper.

Some fish can be overbearing, but we cut some slack on a devil scorpionfish; he must be so patient for snax to cruise by. He gets a bad rap on his venomous dorsal spine too, but if you lived in sandy rubble where others snubbed you or—injury to ignominy—stepped on you, wouldn't you want a toxic spike? Stepping on Scorpio will ruin your day.

Hey, now somebody get in here!

Most land residents of a reef community will defend reef residents from aquarium marauders, preferring abundant reefs over a greedy few making monthly payments. Our 40th POTUS Ronald W. Reagan put it best: “Mr. Gorbachev, tear *down* this aquarium!”

The aquarium trade took hermits by the hundreds of thousands, leaving reefs vulnerable to collapse without this lynchpin species. A hermit needs many shells during its life, yet each hermit taken removed a shell to further weaken the balance.

Just that quickly, we arrive at the far wall of the yonder pinnacle and yet another micro-society, this one less grumpy and more sublime.

Frogfish, in which the piehole is window to the soul.

What's sublime about a frogfish, who may die trying to engulf something larger than himself? For starters, you can meet him nose to nose and not see him. Yet you sense a tiny wheel whirring inside, measuring YOU for a big gulp. He seems American, ready for more. This baby frog could fit on your pinky fingernail—while pondering your thumb.

Benthic fish live on the bottom. Pelagics are free swimmers who might favor a pinnacle for the cover it can provide.

A tiny juvie wrasse stays close to home.

Left, a juvie shortnose wrasse verging on adolescence is likely to avoid drug abuse, teen pregnancy and grand theft auto. Wilderness neighborhoods are like that, allowing parents more peace of mind—shortnose wrasse teen (above), adult (below).

Psychedelic wrasses are born female. If the male departs, even around the corner for a 6-pack or some smokes, the dominant female will morph to male.

Near another small cave is an old familiar: Rocky is a raccoon butterflyfish happily at it on a slab of USDA prime reef schmutz.

The cave is about 18x18" at the opening and 3' deep with heavy shadows inside, until we intrude to light up another wisp.

A red pipefish makes a living near a fuzzy ceiling. This 1st cozzin to a seahorse has a very similar head and appetite, feeding on tiny plankton.

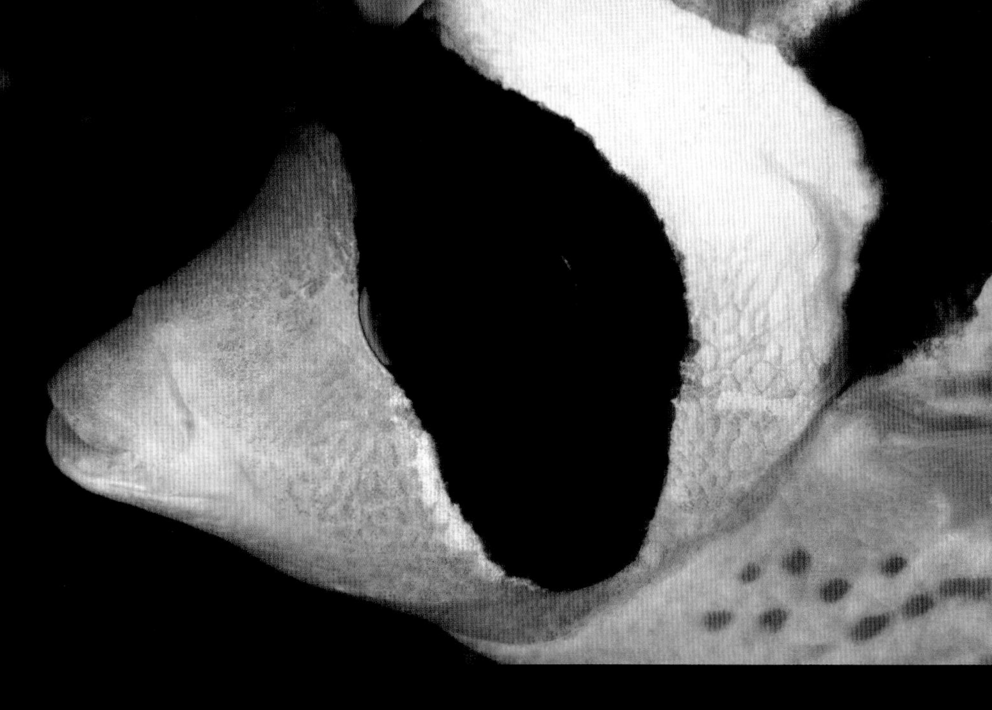

Hey, what's that nibble in my ear?

Portrait of the artist as a young raccoon butterfly

It's Rocky raccoon, whispering that it's time for a lung breather to head back. The current is gaining but not to worry; it's downhill from here and easy, except for the passing parade of perfect pictures presented as part of pattern, in which the waning minutes come alive. Maybe the critters feel more trust after a while. Ah, well, South Maui reefs are gaining abundance and should continue. The home stretch is busy as a receiving line with shots on the fly and air waning.

The primordial ooze, aka a little gob o'snot, may be the origin of more than some lung breathers care to admit. The little gob here is flecked with embryos that will evolve, given adequate generations, into LA commuters.

A flying gurnard awaits tower clearance. We're low on air with no time to dawdle, except for one little frontal—hey, turn around...

My friend Larry keeps checking to see if I want to smell his butthole. I don't, and he needs to sort that out. I suspect sexual identity issues—or that he's realizing the full potential of adolescence, a magical, mystical time that some of us never really outgrew.

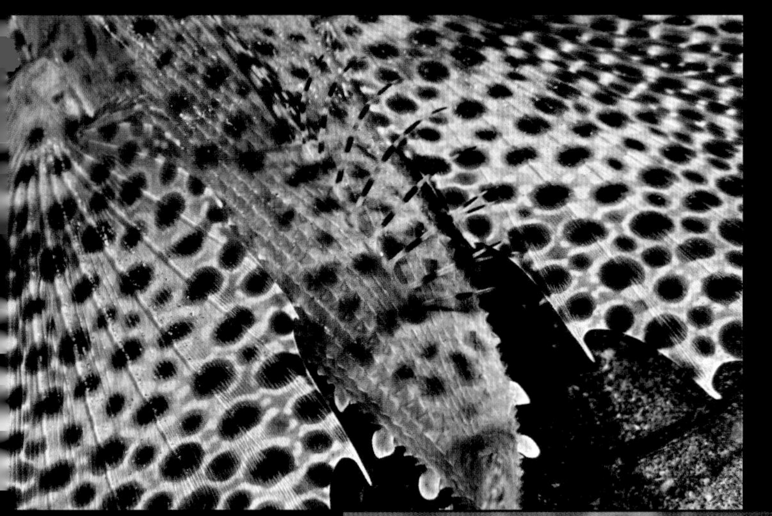

But flying gurnard isn't seeking a sniff test. He's beyond sexual identity or teen pranks. This is life and the converse option on Neptune's way. I drift like flotsam, coming around to the frontal, but he turns away. Why? Defense—check the tail spikes.

Oh, Neptune, it's a baby hermit, so cute! I'd love to stay and chat but….

So many critters come forth in the stretch...

A 4-spot butterfly teen (above) verges on maturity, when all spots will delineate.

Here's how he started, more of a 2-spot with some smudges.

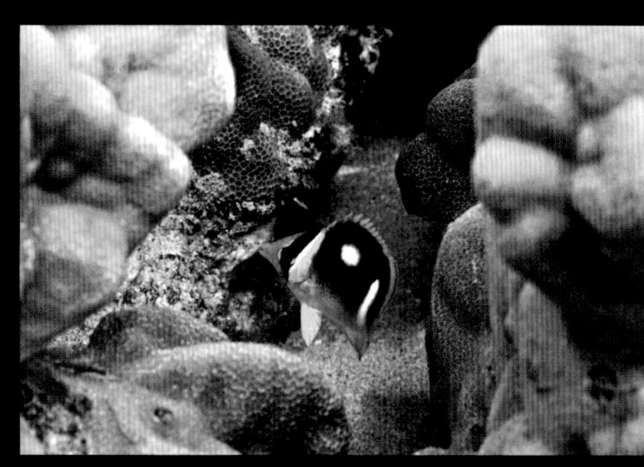

Flagtail tilefish

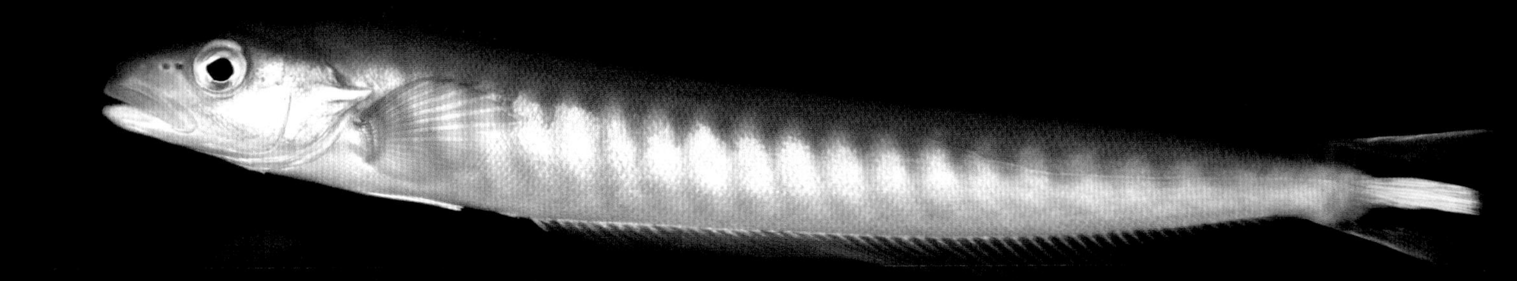

Ho! It's not only 2 adult Hawaiian hogfish.
Look who's hanging nearby!

Hawaiian hogfish adolescent

Hawaiian hogfish adult, female

Baby hogfish makes 3.

Hey, if a network news dog reports on a historical event, I mean an historical event, would she also want to talk about an Hawaiian? Or an hogfish?

Never mind! Dead ahead, as only 400 lbs of honu femme can do, Lisa surges on a single thrust of her mighty flippers! Is this a game of chicken? I've butted heads with some strong-willed womens, but surely Lisa knows that a lung breather in a current can't steer so well. She's gonna ram!...

At the last blink in 1½ knots of current with Lisa bearing down, we both skewed to starboard, sucked it up and nearly rubbed bellies in passing. What a gal—but as my heart went giddy-up the most adorable humuhumubabylei about 6' down on the bottom called up with a question. So I skewed on into a peel-away descent to square with baby lei and…

And then we were out of time! Safe diving means you're up with 500 lbs left, and there I was at 515 on a 55' bottom—with a 3-minute safety stop at 15' to go.

Then, it happened: like a knock, knock, knocking on heaven's door came 3 taps on my back. A dive buddy would grab my fin, I think. And a fish in a gray suit wouldn't be so formal but would dive in to lunch. So I rolled over, camera poised, to see the look… of love… is in… your eyes. The look… your heart can't disguise…

Are you my daddy?

I felt that this was it: she wanted to mate, given my virile profile. Maybe she liked my wrinkly neck. It happened before in *Some Fishes I Have Known*, when another honu femme in XXXL cruised up below me, apparently receptive. Sure, it's fun to speculate on all-star offspring with wit, language skills, reef love and oh, the flippers! I was flattered, nay, honored, but my love for Honu Lisa shall remain spiritual, as true love must be.

Neptune's way hardly fits in a chapter, but this sample of abundance 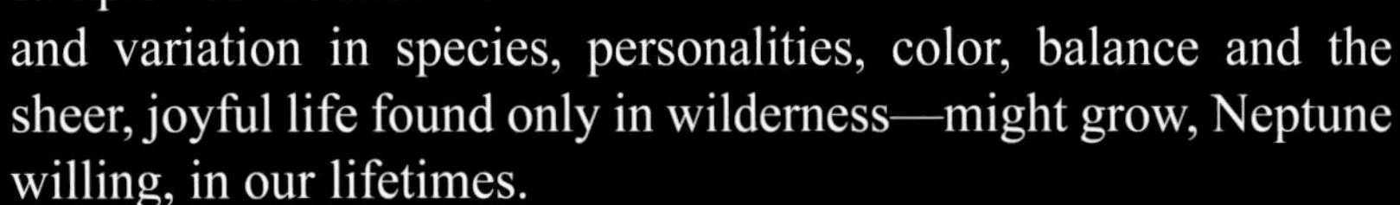 and variation in species, personalities, color, balance and the sheer, joyful life found only in wilderness—might grow, Neptune willing, in our lifetimes.

Wilderness is not compromised by humans coming in, only by humans taking away.

ON COLOR & FORM ADDING UP TO GRACE

The age-old question persists: if an aquarium hunter voids a reef of fish, and nobody sees the event, were the fish ever really there?

This is not a trick question. Orwellian voices may shout down the truth with urgency to make money. One aquarium hunter told the Hawaii State legislature that he could barely make payments on his boat, motor, trailer, disposable diapers for 3 babies and a 4th on the way. With this dire need for a washing machine, a curb on aquarium collecting would drown him, so to speak. Another aquarium hunter testified that puffers are found on nearly every reef, so why shouldn't they be caught and sold? The legislature deferred to these tearful needs.

"We find puffers on almost every reef. So why shouldn't we take them?" Aquarium hunter testimony to the Hawaii State Legislature. Puffers bring $3, then they die.

Yes, I know: this segment of Neptune's speech is named for beauty, but we have some ugly in the water. The aquarium trade claims "sustainability" as it takes everything in reach. Petco is a chain with 950 stores, some trafficking in wildlife. Customer Service on the Petco website goes to the Philippines, where inquiries are redirected, noted and forgotten. Petco labels its marine wildlife as "tank raised." That means it was caught in the wild, very young. A Petco manager explained: "tank raised" is the same as captive bred. When I, Snorkel Bob, assured her that it's not the same, she said, "Well, to tell you the truth, Corporate is working to stop the reef fish sales. Nearly all of them die before we can sell them anyway. And good riddance. I hate it." Alas, Petco still sells reef fish.

The aquarium scourge in Hawaii rails that it doesn't interfere with tourists; it hunts away from the tourist reefs, and kidnapping is okay if you can't see it. The scourge claims that a million fish here or there is nothing. But a million fish is something; you *can* see it. One collector complained that once he filled an order, he had to go ahead and take everything. He didn't want to, because he's a good guy after all. He had to take everything because the bad guys would take them anyway, and who wants that?

Light and color capture a blackside hawkfish ready to swoop.

Sea Shepherd commodore Paul Watson years ago told me how the Sirenian, a high-speed vessel retained by the Ecuadorian government, apprehended a freighter poaching sharks and turtles in the Galapagos.

He was excited. I didn't have the heart to tell him that one less poacher would be a drop in the bucket. Then I lost more heart and more again, till realizing that we lose until the drops add up, one poacher at a time. It's a whimper or a splash, boys & girls.

I'll take a splash, please.
One less poacher does make a difference.

Adult eels are in demand across Asia and New York, where huge home aquariums compensate otherwise arid life. Aquarium extraction is bad policy for anyplace blessed with reefs. Pono is Hawaiian for doing the right thing. The aquarium scourge is not pono. Must I, Snorkel Bob, actually tell you this? Well, I suppose I must, if you live in a parochial society like that found in Manhattan.

Whitemouth moray eels. The scourge drop 3' lengths of plastic pipe over the side, baited & shut at one end to shanghai eels out of Hawaii. It's a crime against nature.

Whitemouth moray eels are common in Hawaii. Tiger moray eels are rare. So a tiger moray eel's address cannot be disclosed lest the aquarium junta take them away. Tiger eels are slow, relaxed critters who enjoy a bottom snooze and don't mind a little chin schmutz, which strikes a home chord with me, Snorkel Bob.

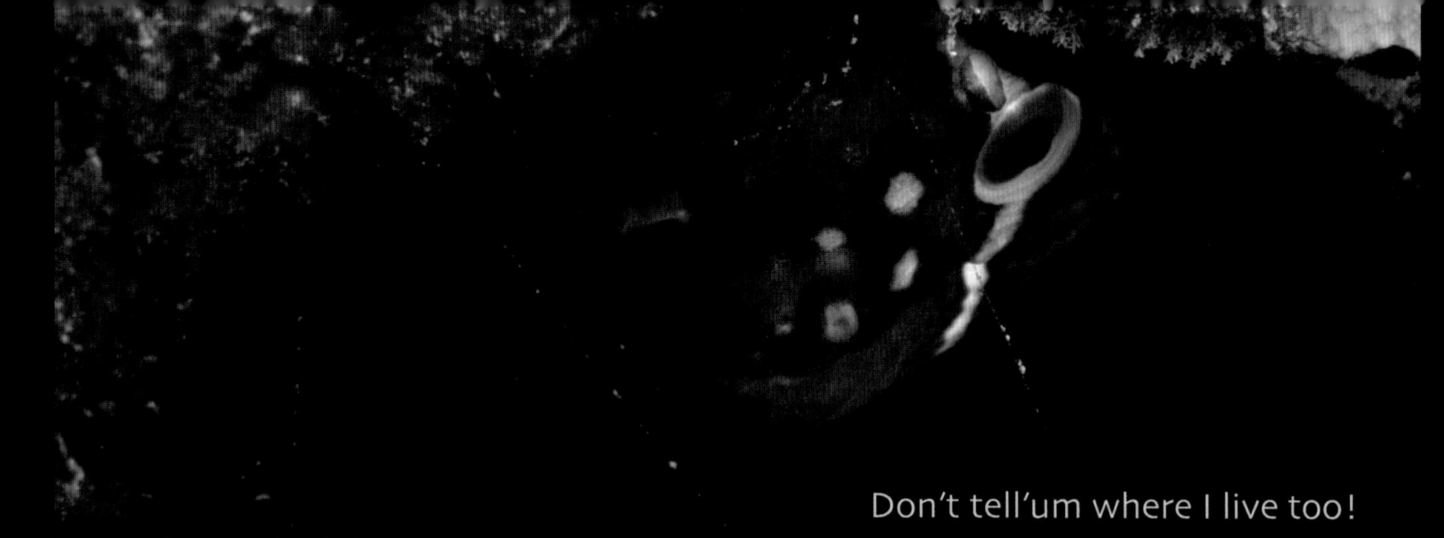

A juvie zebra eel pleads for discretion.

Color, form and grace are sweet as candy spiced hot to open our eyes and ears and make our head sweat. The world will not end on a certain Sunday predicted by kooky Christians claiming rapture for those who love Jesus and hellfire for the rest.

The black durgon is a triggerfish in purple and yellow. Neptune's code translated: the world is ending now as it always has with one fatal difference—recovery no longer keeps pace.

Speaking of grace and wilderness, across from Maui on Lanai are reefs still abundant with species. A Maui aquarium hunter discovered these reefs just prior to Maui County's ordinances against the aquarium trade—that guy moved to Kona to secure his future of wildlife trafficking for the pet trade. He left behind this bandit angelfish.

Another fellow calling himself **The Reef Raper** sells bandits on-line for $400 to $600 each.

Maybe it's not an aquarium hunter but a bandit angelfish who is cast in the image of God. The angel above lives on reef called Cathedrals, where clear, blue water should be rapturous and free as a matter of daily living. Shouldn't it?

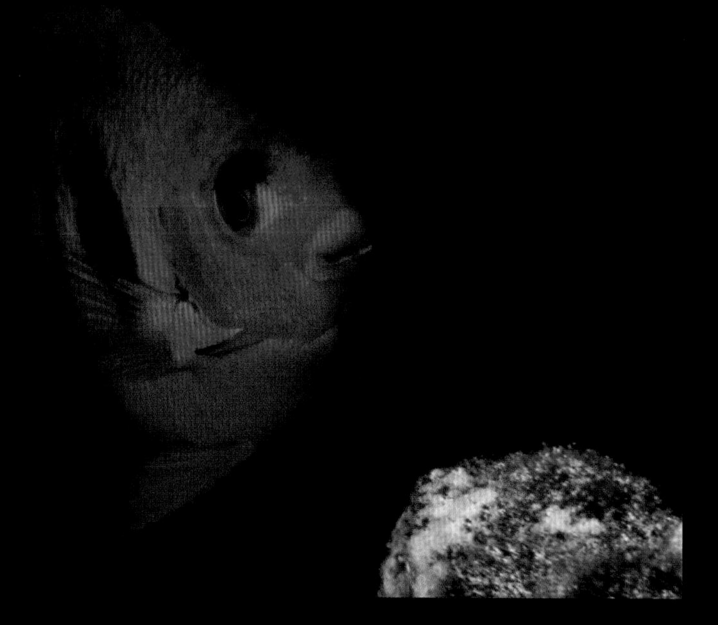

Another blessed place to be is eye level with the angels, beginning with a flame angel, among the rarest on Hawaii reefs. Nobody says where a flame lives; the scourge will scoop her and him and all the juvies in a heartbeat, converting color, form and grace to a few dollars soon spent.

So the man submerges with a smile that can't help but leak seawater and the taste of home—sure, it was the scene of domestic strife in recent days and nights, but that hardly means the family must remain dysfunctional.

Free diving beyond the reef to twenty feet, he rises slowly, blowing his biggest bubble with his head in the center. Into the thickening mix of emperor and regal angels, blue damsels and turquoise chromis, all the little fins and his heart aflutter till—wait!

Flame angels!

Just there—a mated pair, peeking out and darting to other cover, demure as a deb, brazen as a beau and far more virtuous than either. But despite these red-orange bodies and vertical black bars in a riotous fire burst stumbled onto like lost treasure, an unfortunate segment of humanity wants to capture and contain them, wants to watch them under control, in captivity, as they fade away.

—from Flame Angels, a novel of Oceania

A bicolor angel (above) is shy as any and darts for coral cover to take a powder.

Angelfish are everywhere, with grace to share, like these angels in Fiji.

Also shy, a 2-spined angel is similar to the Potter's angel in Hawaii.

Potter's angelfish in Hawaii bring news from Neptune: time is running out.

It's plain to see that Neptune ain't whistlin' Dixie on grace and form following function. Have you ever seen such extravagance, even at the mall during Chanuka?

Reef Hole Garden Party is a gathering of reef people working their assigned tasks. The Hawaii cleaner wrasse cleans angelfish and many other species severely depleted by the aquarium scourge—species who live to graze and thereby balance the world's reefs.

A yellowtail coris poses the painfully obvious question:

I get it. Do you get it?

A 4-spot butterfly grazes primarily on cauliflower coral polyps.

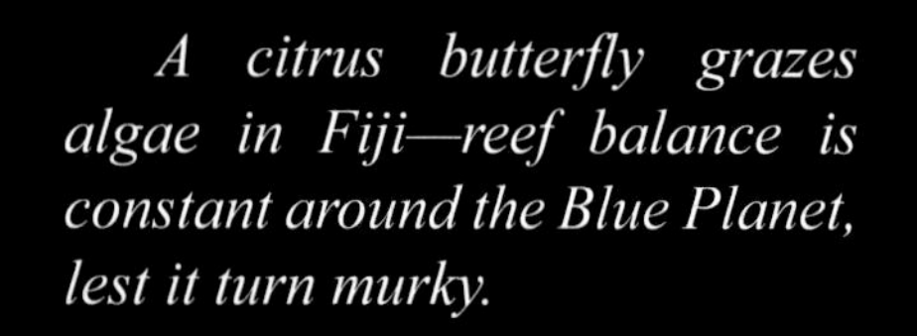

A citrus butterfly grazes algae in Fiji—reef balance is constant around the Blue Planet, lest it turn murky.

If not grazed, algae will overtake coral. A goldring surgeon, or kole (koh·leh), is an herbivore like the yellow tang, controlling algae on Hawaii reefs, yet she is taken with no limit by the aquarium trade, and so are...

...Hawaiian cleaner wrasses, found nowhere else in the world and critical to reef health. Cleaners pluck parasites & goobers from skin, scales, gill plates and choppers of every vertebrate reef species. Remove the Hawaiian cleaner wrasse from the love scene playing out here, and what do you have?
Pucker up, Shweetheart.
Touch me here, fool.
Oh, Neptune, don't stop.

Convict tangs, or manini, recently schooled in the hundreds on many Hawaii reefs, yet they too are decimated. With no herbivores, algae takes over.

Oh, you'll get it one day.

An oval butterflyfish (above) also favors cauliflower coral polyps. A forceps butterflyfish (below) seeks small inverts.

Schmutzing is also a vital reef function.

Manybar goatfish (above) often sifts the bottom to keep the process moving, as an orangespine surgeon (right) grazes coralline algae nearby.

A yellowtail filefish grazes coralline algae.

A goldtrim surgeon grazes coralline algae, while a parrotfish exposes new substrate in accordance with Neptune's plan. Parrotfish are a lynchpin species whose absence puts a reef in peril. The State of Hawaii condones total extraction at night by speargunners on compressed or surface-supplied air, while the parrots sleep.

Forceps butterflyfish (right) and yellow tangs graze coralline algae from dawn to dusk and support a marine tourism economy as well, yet they comprise 65-80% of the aquarium catch in Hawaii, leaving by the millions, leaving reefs vulnerable to algae overload….

…which brings us to Neptune's paradox on sheer, crazy love…

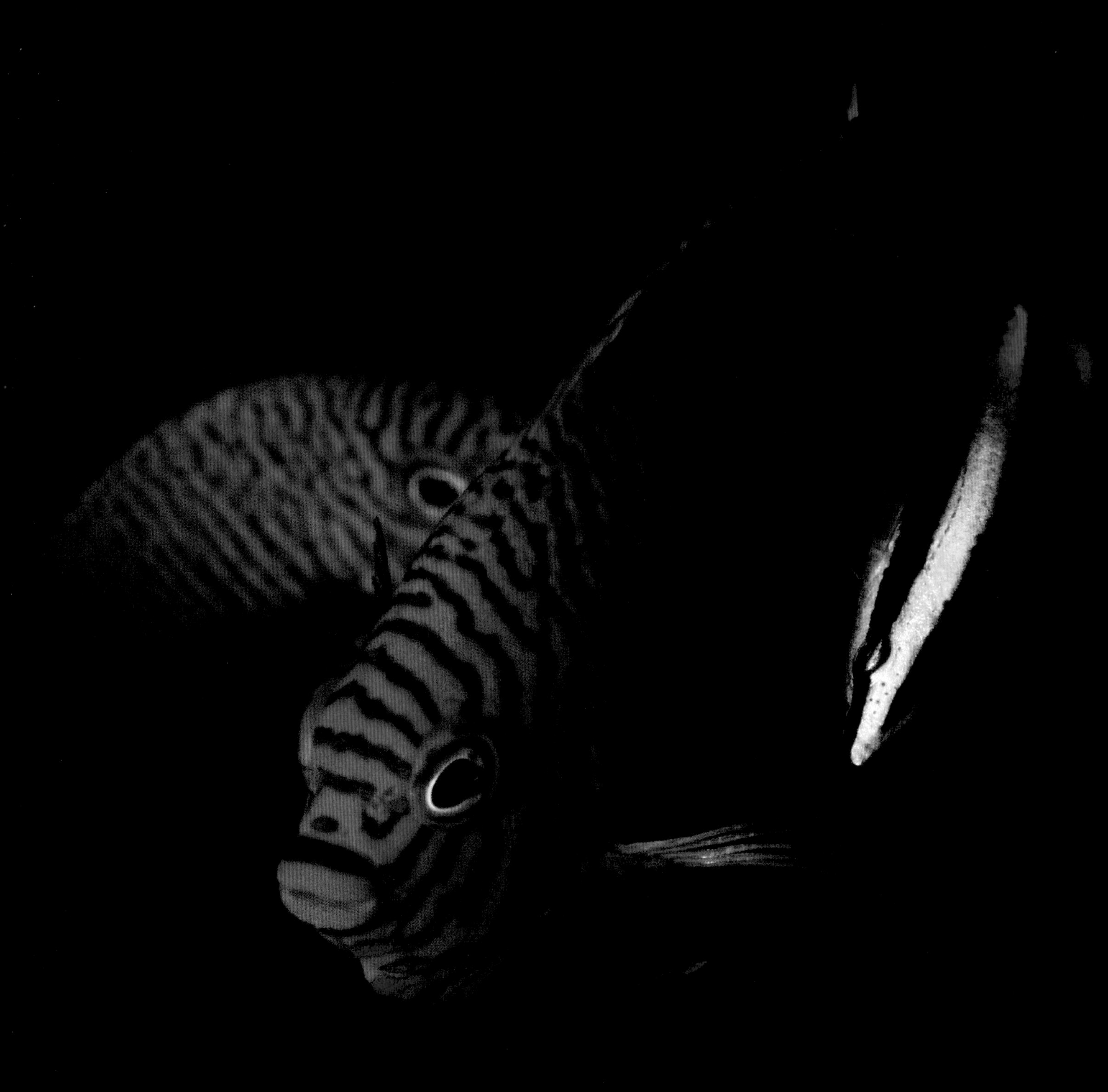

Neptune takes a holiday on certain moments of no material value except for the highest point of life itself, which is sheer, crazy love:

Mr. & Ms. Elegant Coris at home, North shore, Maui

I can hear her heartbeat for a thousand miles
And the heavens open every time she smiles
And when I come to her that's where I belong
Yet I'm running to her like a river's song

She give me love, love, love, love, crazy love...

—Van Morrison

Well, how far can you go on this raw beauty business? Forever? Because I got the tread to get us there, but that might spell the difference between crazy love and just plain crazy. So we'll split the difference for a once around the block, just for fun—and love.

Joie de vivre in soft coral, a baby wrasse...

... and a *Scalefin anthius, Fiji (right)*

Fiji: (top) tiny fish in the ½" range cruising soft coral. (Bottom) triangular butterflyfish, relaxing on the set, between takes.

(Above) Scalefin anthius, (below) canary fang blenny

Many humans see themselves in the image of a higher power. Just so, many fish revere the Cosmic Fang Blenny Who demands prayer for personal favors and warns of Her wrath. NOTE: though reverent, most fish will not fight for their beliefs. The fang faction may requisition its bit of flesh, which is different than inquisition.

Actually, reflective flecks surround this canary fang blenny. They look like stellar bodies and may just be.

I, Snorkel Bob, avoid recreational substance abuse, though I do get strung out on anthius in a current, flashing back to Filmore West in '69. Anthii? Anthiuses? Anthium? Anthiosis! *Scalefin (above) hawk anthius (below), Fiji*.

I also suffer loose shutter syndrome on forceps butterflies. It's a compulsion transcending luminescence to embrace flair, drama and sheer crazy love…

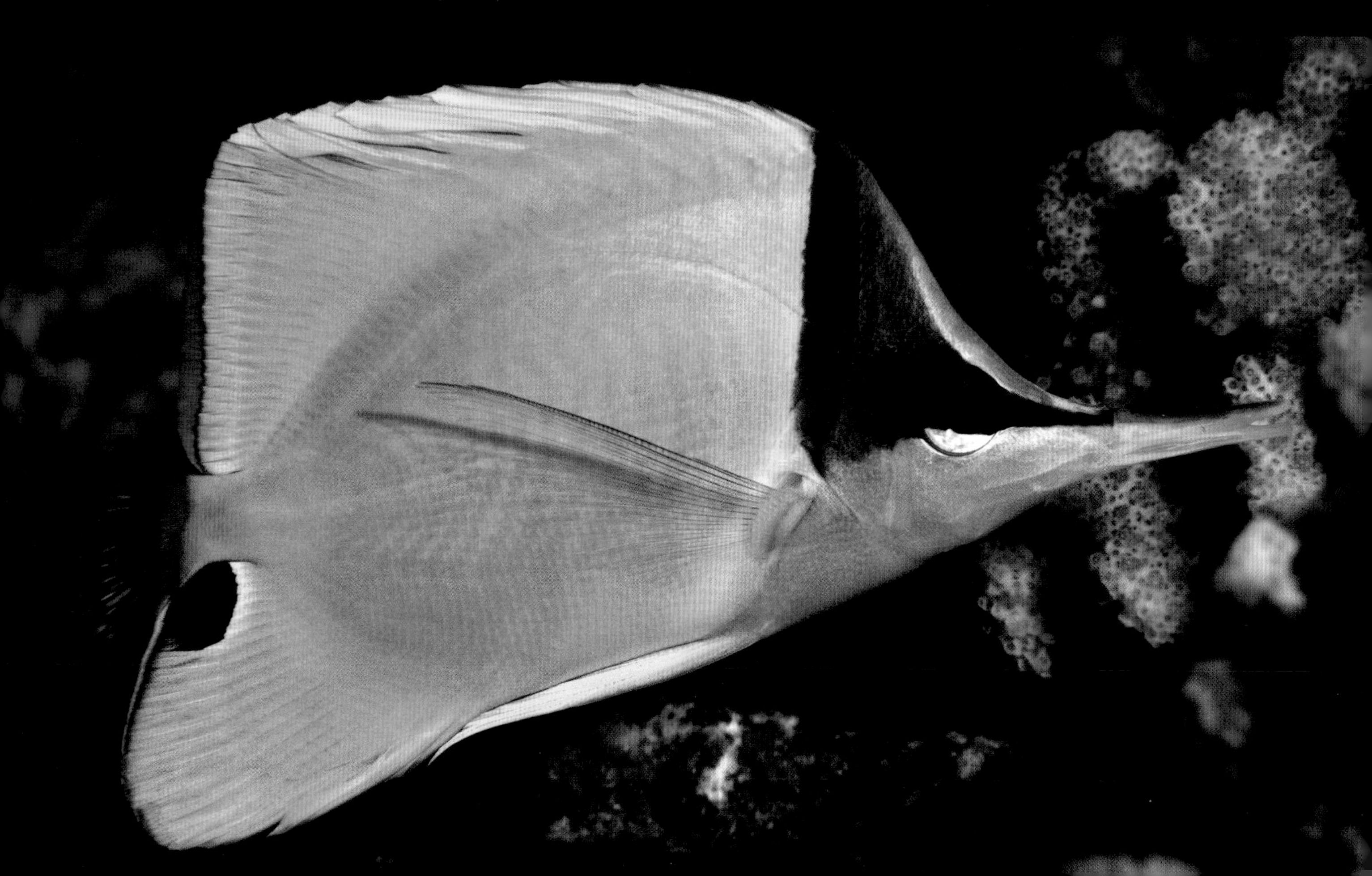

The longnose butterflyfish below has a longer snout and higher forehead than the forceps butterfly above. The key longnose trait: freckles under the gill plates:

Like me, Snorkel Bob, the longnose butterfly has a dark phase that can be beautiful. *The longnose butterfly (above) is yellowish, indicating the beginning or end of the dark phase.*

ON MORE CONVENTIONAL ROMANCE INVOLVING MUTUAL NEEDS & CONVENIENCE

Gill breathers can bond. Some fish are monogamous in the pure sense, mating once for life. If a mate dies, the survivor remains single. You could make a case that the mating drive is genetically programmed to occur early and briefly. Or that pure monogamy is a fishy vow, like the celibacy some human sects practice—though fish rarely molest their male juvies.

Besides a greater common sense and grace, reef residents seem more naturally attuned to their behaviors. These bluestreak gobies assess the big lug approaching:

Bluestreak gobies are common to tropical latitudes. These 2 live at 65' near Taveuni Island, Fiji. I, Snorkel Bob, have seen them on Maui in a mall pet shop @ $15, looking frail and fin-rotted from dirty water in a dirty scene. ("You're the guy tryin'a put me outa bidness!" What a piece of work is man.)

Indigo dartfish, left, and chromis, below, enjoy the company of peers.

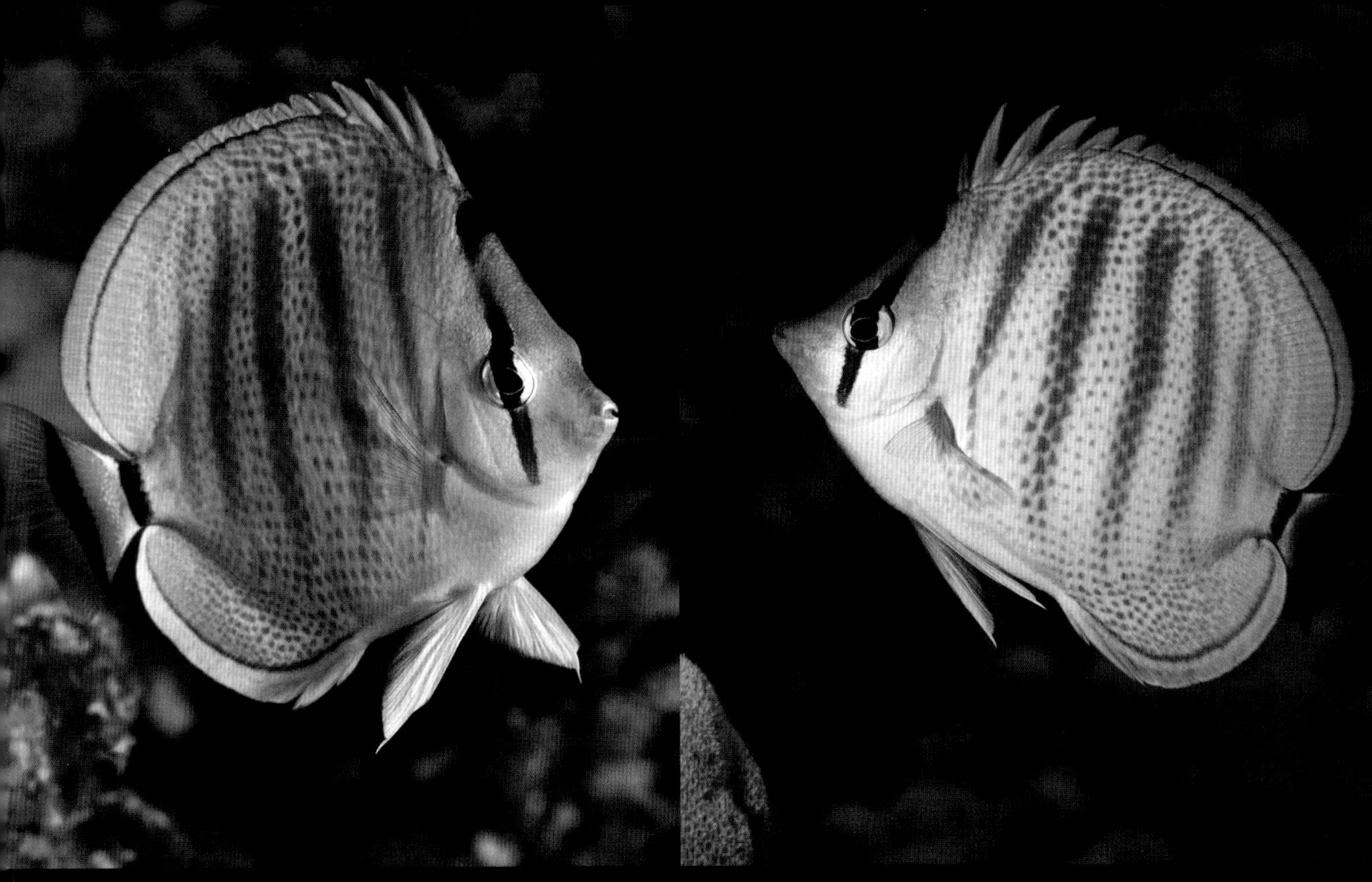

Butterflyfish are monogamous and seldom seen without a mate. Multiband butterflyfish (above), raccoon butterflyfish (below).

Ornate butterflyfish (above), Blueline butterfly (below)

Dot & dash butterfly, Fiji

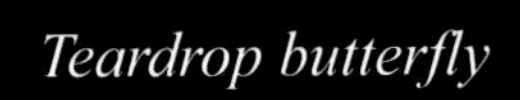

Teardrop butterfly

Milletseed butterfly

Blacklip butterfly

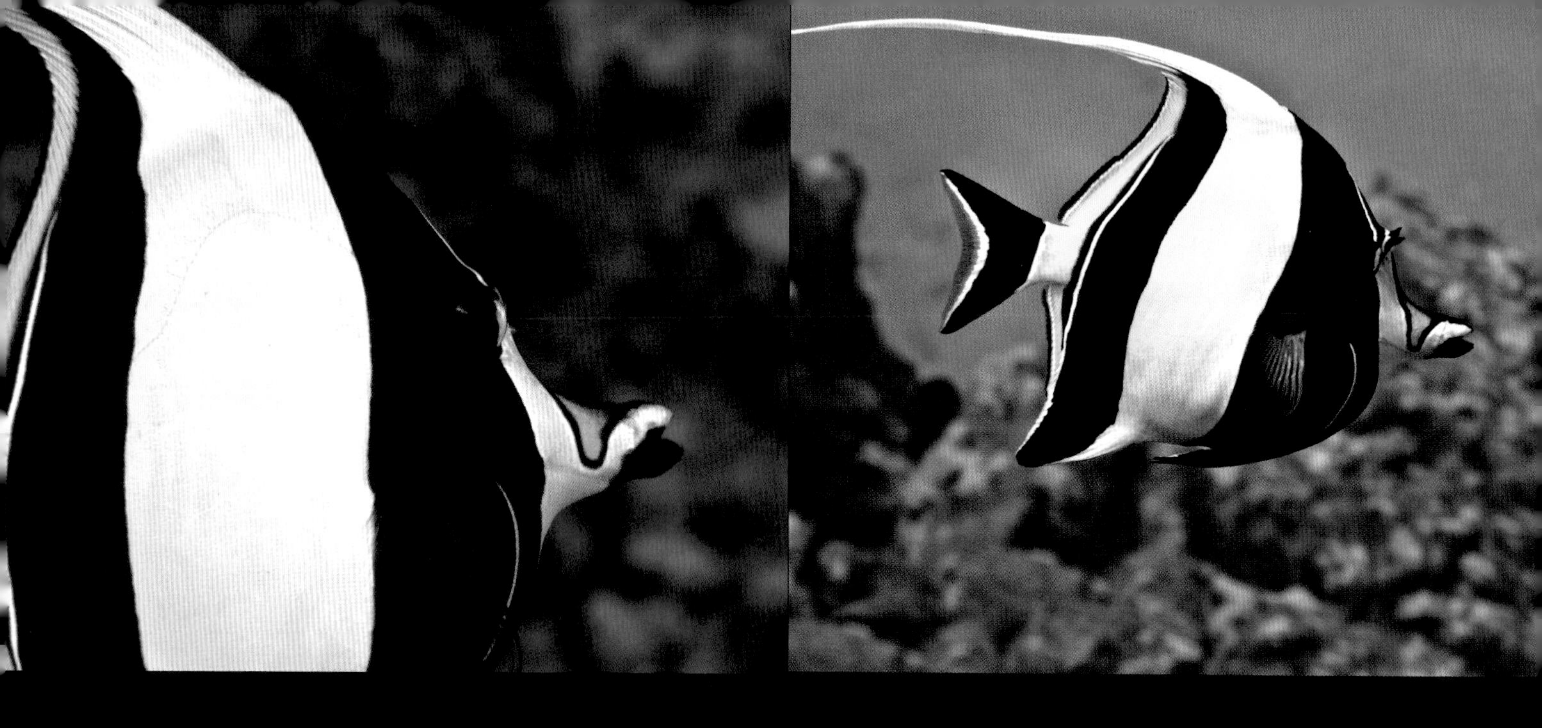

Moorish idols are a species unto themselves. Like butterflyfish, they are monogamous and seldom seen without a mate.

Bonded pairs abound, like these blue chromis (right) and anemone clownfish (below). Neither species appears to suffer from the quirks or petty habits of the mate.

But both species suffer severe pressure from the aquarium trade.

Friends of different species bond too:

I don't care who thinks what; he's young & attentive, and he doesn't get tired.

I don't mind helping you out. I just don't want to hurt Steve...

I, Snorkel Bob, get reef porn on my tweeter all the time—ill advised and crude, it also shows bad judgment, especially if a fish wants to run for Congress.

With Parental Guidance at hand, this may be the time to assure that somebody is out there for everyone, and even cowries have foreplay.

Some people are naturally social, like cleaner wrasses, setting up cleaning stations where friends gather. The cleaning station may be the most social part of a reef.

Nudibranchs may be more practical than romantic, but who knows? *Imperial nudibranchs, Maui.*

Reef society encompasses girlfriends…to roommates…to whip coral mates who meet surreptitiously…

Some bonding is practical and touching. The roommates above are a shrimp goby who guards and snatches

snax to share with the goby shrimp, who is blind and does the housekeeping and keeps an antennae on the goby.

Tiny rainbows below burst from a coral head in Fiji then retreat to its cover and repeat as necessary.

Some people don't do well alone, like this remora or pilot fish, who might cruise solo but prefers glomming to

a bigger critter—shark, barracuda, turtle—as if helping to lead the way, gaining protection and the odd crumb.

Schooling is considered a protective behavior, but it's also a social interaction.

Hawaiian squirrelfish cruise casually, (left).

How it used to be and maybe can again, at least on Maui, the only island in Hawaii to ban gillnets and turn the screws on the aquarium trade that would take every one of these yellow tangs in a fell swoop and call it sustainable. Why shouldn't they?

Besides bonded pairs and odd friendships, many

Hawaiian dascyllus or damselfish share a coral head and scurry for cover if a threat approaches, yet they ease on out if he's not predatory or has no net.

Aquarium keeping is cruel and unusual, directing home hobbyists to run-in their tanks with "cheap" fish. Bad chemistry kills fish, so running-in with dascyllus is "smart;" plenty more where they came from.

Would they recommend LA commuters as an expendable run-in species?

Below: Hawaiian dascyllus at Honaunau, South Kona, and (above), an orangefin dascyllus over antler coral in Fiji.

A tiny pup dascyllus on Maui and a dascyllus in Fiji are fair game for an unlimited aquarium trade calling these creatures expendable.

Which raises the big question on the aquarium trade or any threat spreading across the land, leaving a wake of shame, shame, shame.

Why don't we put things to right, for the children?

A juvie Hawaiian lionfish meditates on reef abundance and liberation even at his tender age.

ON DOING IT FOR THE CHILDREN

I, Snorkel Bob'm not suggesting that we do *it* for the children; it would be, as they say, inappropriate, and they would not appreciate it, most of them.

But—and it's a big but—if we don't protect our reefs for generations of fishes to come, who will?

They're so cute at that age, once they get their dimples and cold n' fuzzy little puppy dog faces. Have you spent time in the nursery?

You can't imagine the stress and strain old Neptune endures—at his age (!)—till you swim a mile in his fins. But who's complaining, with cuddles abounding? *Juvie damsel taking comfort in anemone tentacles, Fiji (below).*

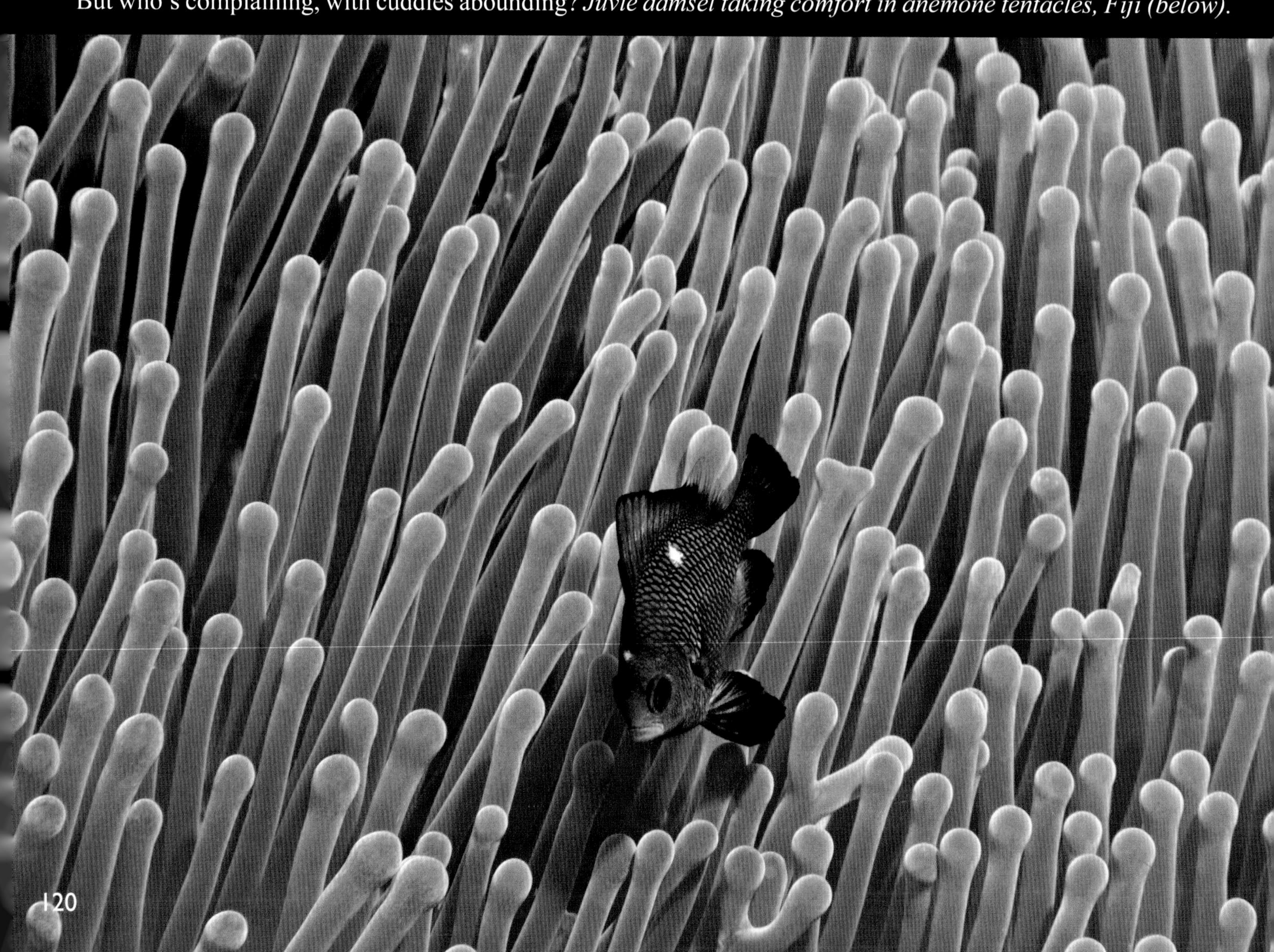

Squint on in to this anemone clownfish about ¼" across, too young to wonder what life is all about but just the right age to love his anemone home. *Juvie arceye hawkfish (below)*

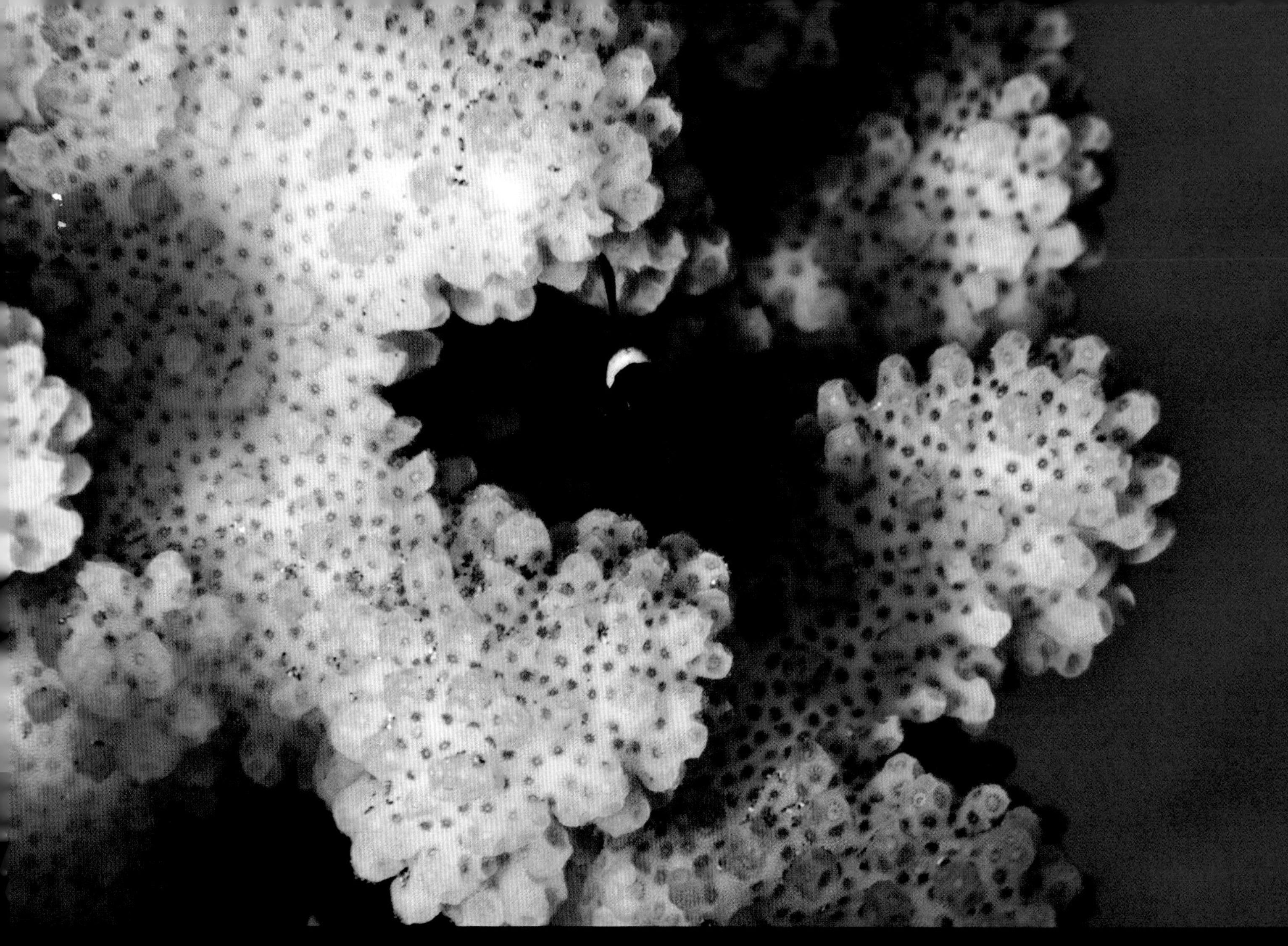

All the wild children: above, a tiny dascyllus. Below left, a southern tubelip wrasse cavorts in coral with a juvie triangle butterflyfish, Fiji. Below right, a young yellowstripe coris, endemic to Hawaii.

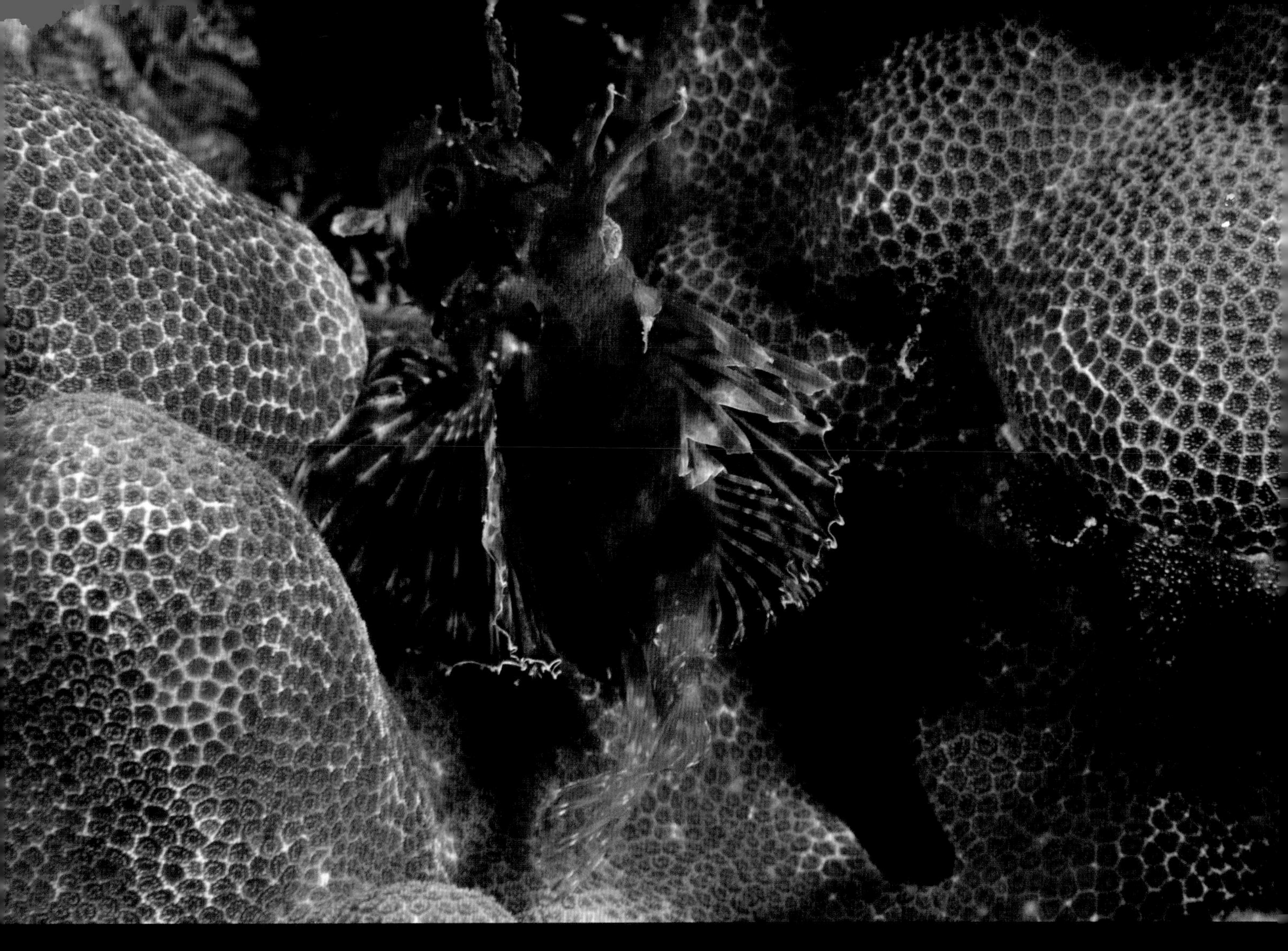

A young lion—Hawaii lionfish, that is—takes cover from predators with teeth or nets. As well as a youthful but circumspect elegant coris (below left). And so does a juvie speckled scorpionfish (below right).

Who'd a thunk a cuddly oval chromis juvie above could swell and fade as many people do, to the adult, below?

A juvie psychedelic wrasse female

Left middle, a juvie 2-spot wrasse

Right middle, a tiny eyebar goby hardly 1" long

Bottom, a juvie puffer in soft coral, Fiji

No juvie? Look above the clownfish piehole and dorsal fin. Below, a juvie striped wrasse on a Fiji reef enjoys youth just like…

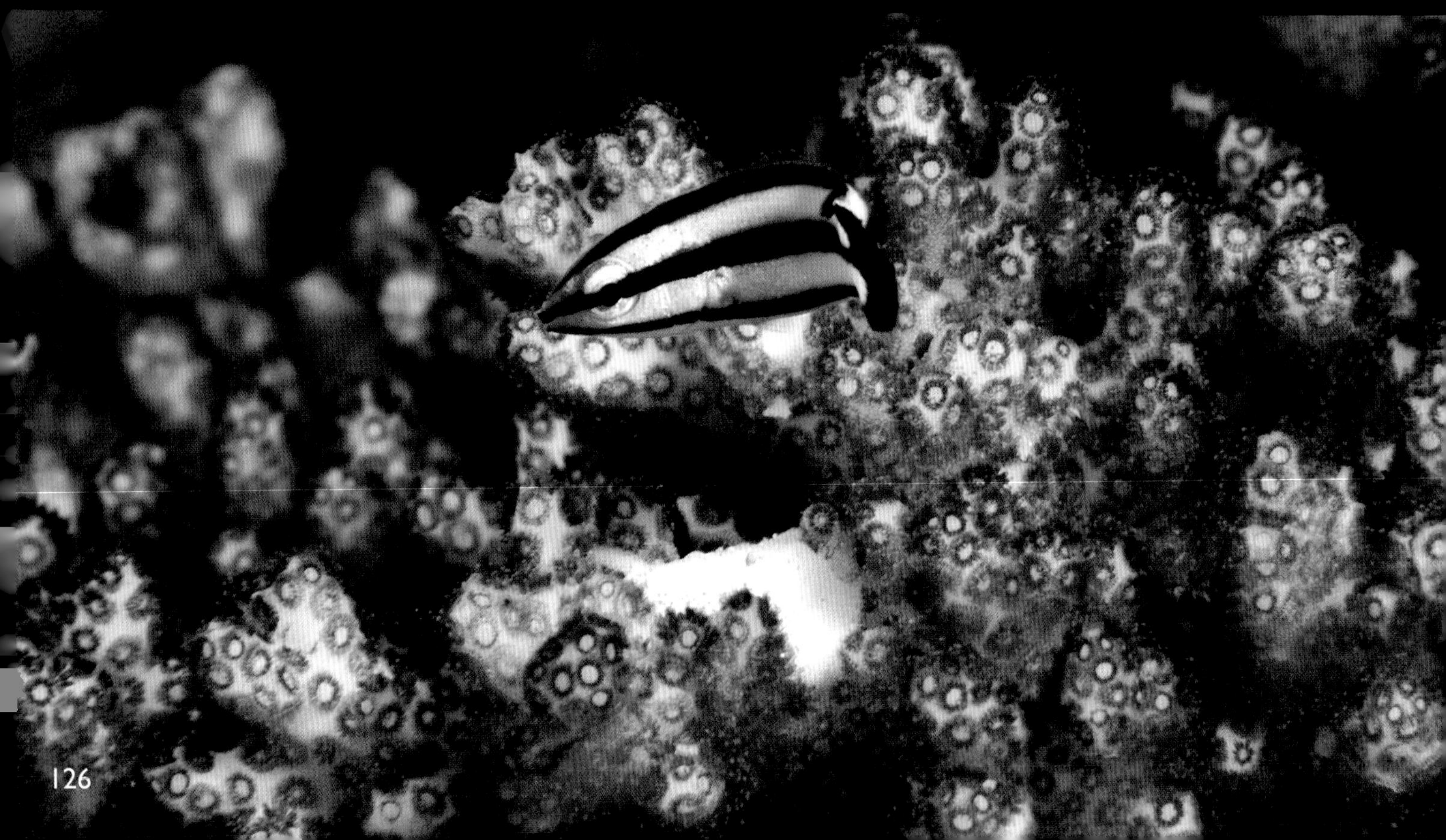

...juvie yellowtail coris on Maui reefs. A young frogfish at only 2½" looks presidential, ready for Mt. Frogmore.

A shortnose wrasse and elegant coris browse for snax, above. Middle, a slightly older shortnose poses for the camera and, bottom, a young chocolate dip chromis displays radiant blue trim.

Top, a baby razor wrasse. Middle, a young blackbar razor wrasse.

Check out about ½" below this baby razor's chinny chin chin to find a much smaller baby that I, Snorkel Bob'm advised is a tiny goatfish, who with luck will become:

As the only island in Hawaii with a gill net ban, Maui may be the only island with a resurgence of goatfish and parrotfish too—and unicornfish and yellow tangs. Sadly, the State of Hawaii still allows night spearing of parrotfish on scuba or compressed air, blighting the future of parrot children and children everywhere.

The Cruise Brothers:

On a happier note, aquarium hunters and dealers are fleeing Maui, allowing a reef resurgence. Young tangs beyond recruitment size may now reach spawning age.

ON FANG BLENNIES, THE GREGARIOUS RAPSCALLIONS

Fang blennies are not cleaners. They set up fake cleaning stations then dart in to feed on the scales, skin and snot of other fish—and/or snorkelers and divers. The saber-tooth fangs on the lower jaw are not used in feeding but in defense, and yes, these little buggers can move on an unsuspecting waterdog.

Here's the rub: any guardian will attest to loving all wards equally, no matter their mischief. But Neptune has his human frailty too and seems especially fond of fang blennies, with their happy faces and dastardly tricks…

Left, piano fang blenny

Below, ewa fang blenny

Fang blennies vary in color and pattern with a common modus operandi—to flat guarantee a one-owner, low-mileage ride with immaculate maintenance records kept by a teetotaling Sunday school teacher who mostly stayed home to watch PBS and read books. Unless it's a high-mile, off-road, drag chariot run to rattles by a crack addict—BAM!

The ewa (eh·vah) fang blenny is most common in Hawaii (previous page) and finds happiness in Fiji soft coral too, above. Below, Mr. Ewa relaxes at home.

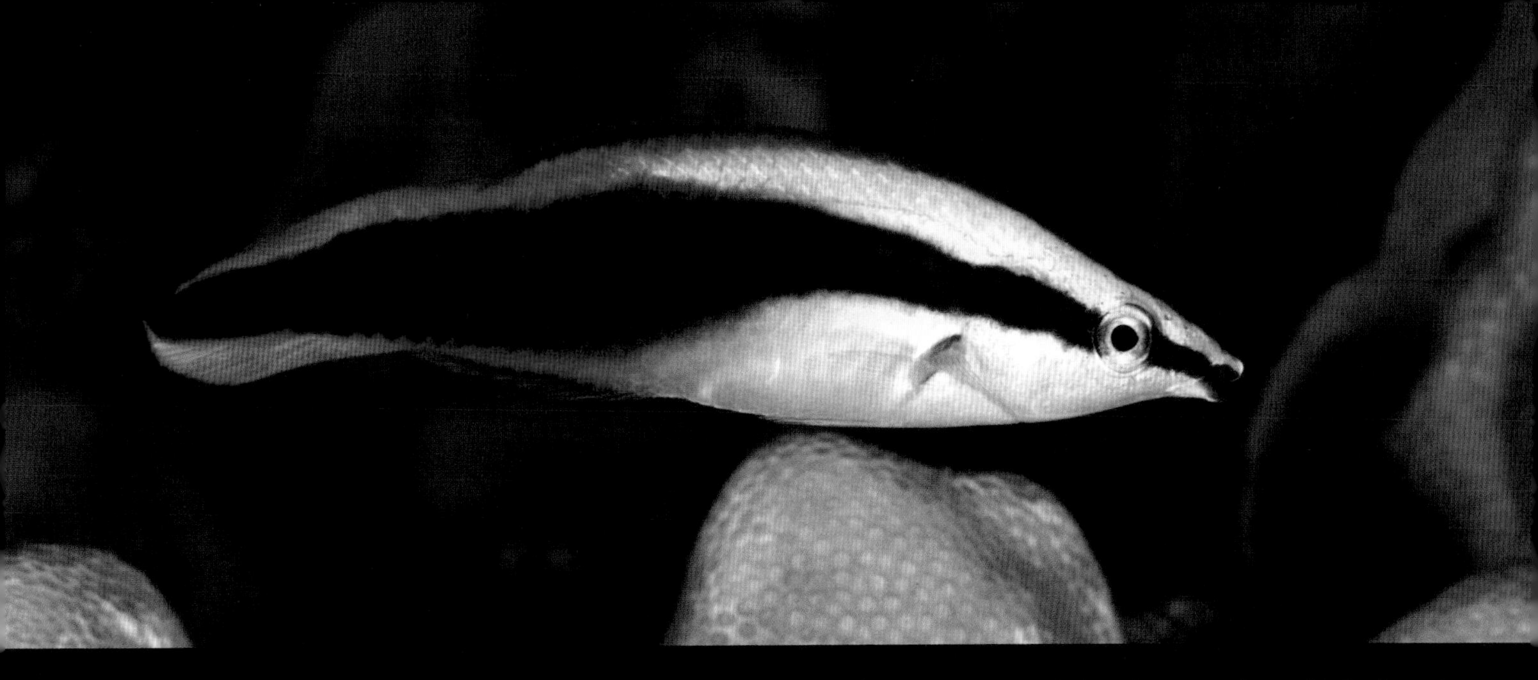

A fang blenny looks nothing at all to me, Snorkel Bob, like the irrepressible, charismatic, widely loved Hawaiian cleaner wrasse or any cleaner wrasse from the Great Reef to Fiji (below).

But worldview differs for gill breathers. Color fades at 40' . Neptune's deft palette becomes shades of gray conveying meaning, caution or aloha. Cleaner wrasses move like all wrasses, with a pectoral fin pump that takes them up a notch on each stroke for a forward glide till the next stroke. Fang blennies move more smoothly and quickly through S-patterns with stalls and face-offs.

Oh, don't touch me there!

Like this ewa fang blenny who was feeling his oats one morning last week, darting at will till everyone on the coral head was astir. Did he give a flat flying fig? No! He darted up with a cherubic grin to see if I, Snorkel Bob, might want to have a nice day.

Well, fool me once, as George Bush once said. But trust is not the same as intelligence; it's often blind. How could we get along in life and on the reef, fearing every proposition? Oh, and that smile (the fish's)…

Note the stoic disposition on this critter. He's not a fang blenny but a curious wormfish (in the dartfish family), cruising the halimeda grass over a sandy bottom.

The fang blenny above is not in my guidebooks—no big deal; those books avoid nuance and character. One author says a million fish mean nothing. He shows dead fish with data on stomach contents of 7 or 19 specimens and tooth counts.

A fang blenny smile will open on formidable fangs. I, Snorkel Bob, don't have that photo, but that guidebook does—the fang is dead on a black backdrop, its smile peeled back to expose the fangs, egregiously colder than the cold-blooded fish. The shot tells more about the author than the fangs—it demonstrates reef love just as gynecology sheds light on romance. Healthy equipment is best, and we are curious. But clinical dissection is not heartfelt. And an old—pardon me—fogey who last dove a decade or 2 ago should beware: flattery from venal interests may cloud a reputation. I ponder age myself and still stick my head in the water to better sense the pickle we're in.

How can I put this tactfully? That fogey is like an old jack offshore who must keep his wits or get gored by a guy with a spear. I know marine scientists who dove with this old man, who saw exotic species and stayed mum, knowing he'd scoop them for data glory on a black backdrop, proving that some scientists actually do feel love. Or a conscience.

The piano fang blenny, right, is uniquely colored with a larger dorsal fin than most fang dudes—yet his charm and ruse are the same.

In fairness to these lovable rapscallions you may want to strangle, I, Snorkel Bob, recall encounters at 80-100' in low light and strong current, where a body senses distance and removal, till a little light cracks the gloom. It comes in with a smile to say *bula*, which is Fijian for…

...Aaaaaaloha.

Waterdogs hear frequencies inaudible to average pedestrians. Neptune's song is a unique greeting; a canary fang blenny indicates that a reef is good. With repetition comes familiarity, and a dialogue develops.

Fang blennies swim freely and are comfortable at greater depth, unlike their cozzin blennies who stay close to bottom cover, sometimes shallow enough for tide wash.

Canary fang blennies, Fiji

A flanneljammies male guards a nest. Formerly known as the shortbody blenny, the flanneljammies blenny changed his name on realizing that shortness is relative in all things and may be lackluster, whereas flanneljammies can be full of potential.

A scarface blenny, below, peers out, ever ready to duck back in, because all blennies like to dive under the covers now and then, like watching TV in bed but with better content.

I, Snorkel Bob, surfed across a Discovery Channel scarefest *Deadly Stripes*, about tiger sharks, where a macho guy swam with a tiger shark. It was deadly, dangerous and vicious: the usual Discovery blather. But the guy befriended the shark, and the narrative factored what she, the shark, saw and felt. Okay, he was no Snorkel F. Bob, but he was very good,—even as the narrative got stuck on deadly hokum, the shark guy was *man enough* to handle the interspecies harmony—beyond the deathly blood & guts blather.

And I wondered: would Discovery make a deadly show about me, Snorkel F. Bob, befriending the most mischievous fish the world has ever known? I feel *man enough*!

Bird wrasse, male

NEPTUNE SPEAKS ON THE WORD

A-well-a, everybody's heard about the bird
Bird, bird, bird, b-bird's the word
A-well-a, bird, bird, bird, b-bird's the word
A-well-a, bird, bird, b-bird's the word
A-well-a, bird, bird, bird, b-bird's the word
A-well-a, don't you know about the bird
Well, everybody's talking about the bird
A-well-a, bird, bird, b-bird's the word
A-well-a, bird
Bbbbbbbbbbbbbbbbb, aaah
Pa-pa-pa-pa-pa-pa-pa-pa-pa-pa-pa-pa-pa-pa-pa-pa
Pa-pa-pa-pa-pa-pa-pa-pa-pa-pa-pa-pa-pa-pa-ooma-mow-mow
Papa-ooma-mow-mow-mow

—The Trashmen, 1963

All birds used to be fish.

—Snorkel F. Bob, the next millennium

Some flanneljammies blennies are believed to have taught university level history in former incarnations.

Okay, boys and girls, it's time for stretching exercises, in which we open our hearts and minds to *see* what Neptune is saying. We won't lip read. We'll freely mix the senses to achieve verbal buoyancy. I.e., underwater you might hear *glub glmphk dufoogleump glumpf*—but it's much clearer than that, if you can see it.

When a shy yellow chromis asks if he may have a non-word, am I, Snorkel Bob to say no? I think best to listen up for Neptune's message. In these pages, a few couriers will bring the word into view.

Gilded or bluegill triggerfish.

Chocolate dip chromis

Whitemouth moray eel

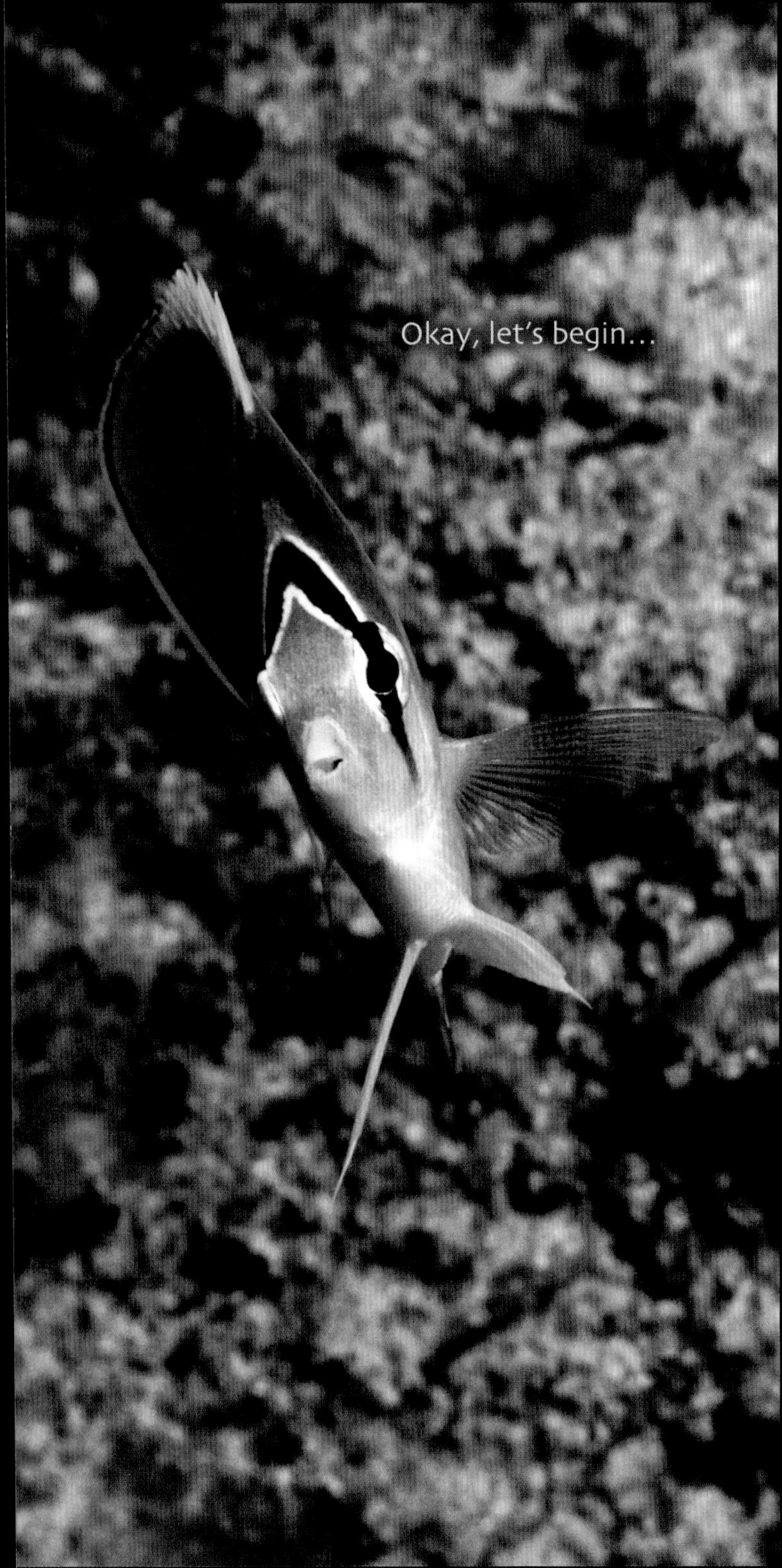
Okay, let's begin…

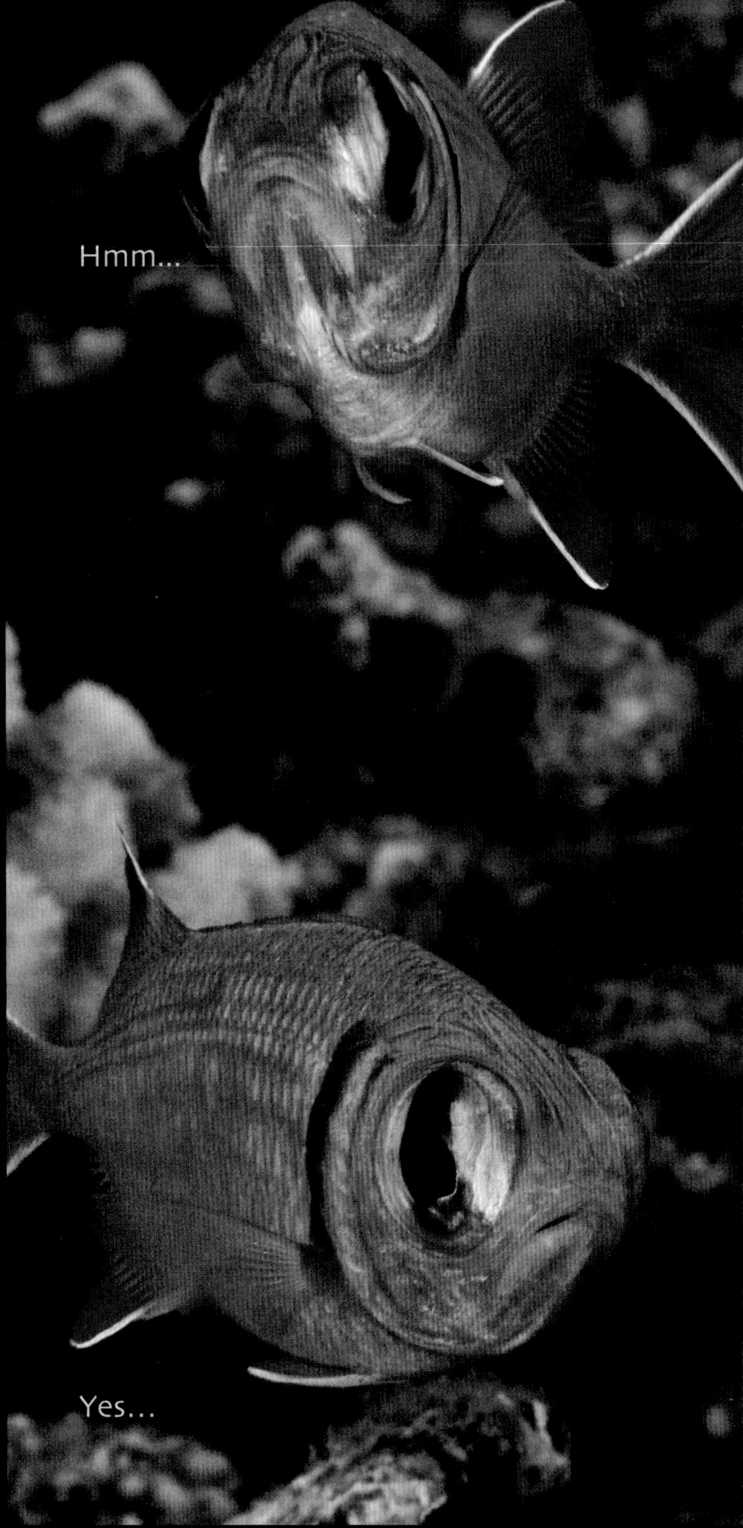
Hmm…
Yes…

A sleek unicornfish in dorsal flair indicates need for parasite removal, yet no Hawaiian cleaner wrasse comes forth—because none are left on this reef.

Hello!

State of Hawaii!

Do you read me?

I'm waiting to be cleaned of parasites—and waiting and waiting, because the aquarium scourge takes all the cleaner wrasses, legally! Are you stoopid? Or is it worse than that?

Ua Mau ke Ea o ka ʻĀina i ka Pono

The life of the land is perpetuated in righteousness.

Above, Ulua or jack trevally

Right, Manini, or convict tang, is a vital herbivore who grazes algae, dawn to dusk.

Below, Oriental sweetlips

3-spot angelfish

Speckled sandperch

Speckled damselfish

Yellow damselfish

Speckled damselfish

**Mr. Gorbachev, tear DOWN this aquarium!*

Blue chromis

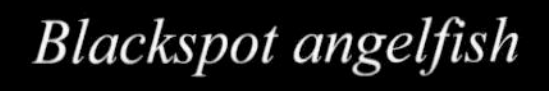

Blackspot angelfish

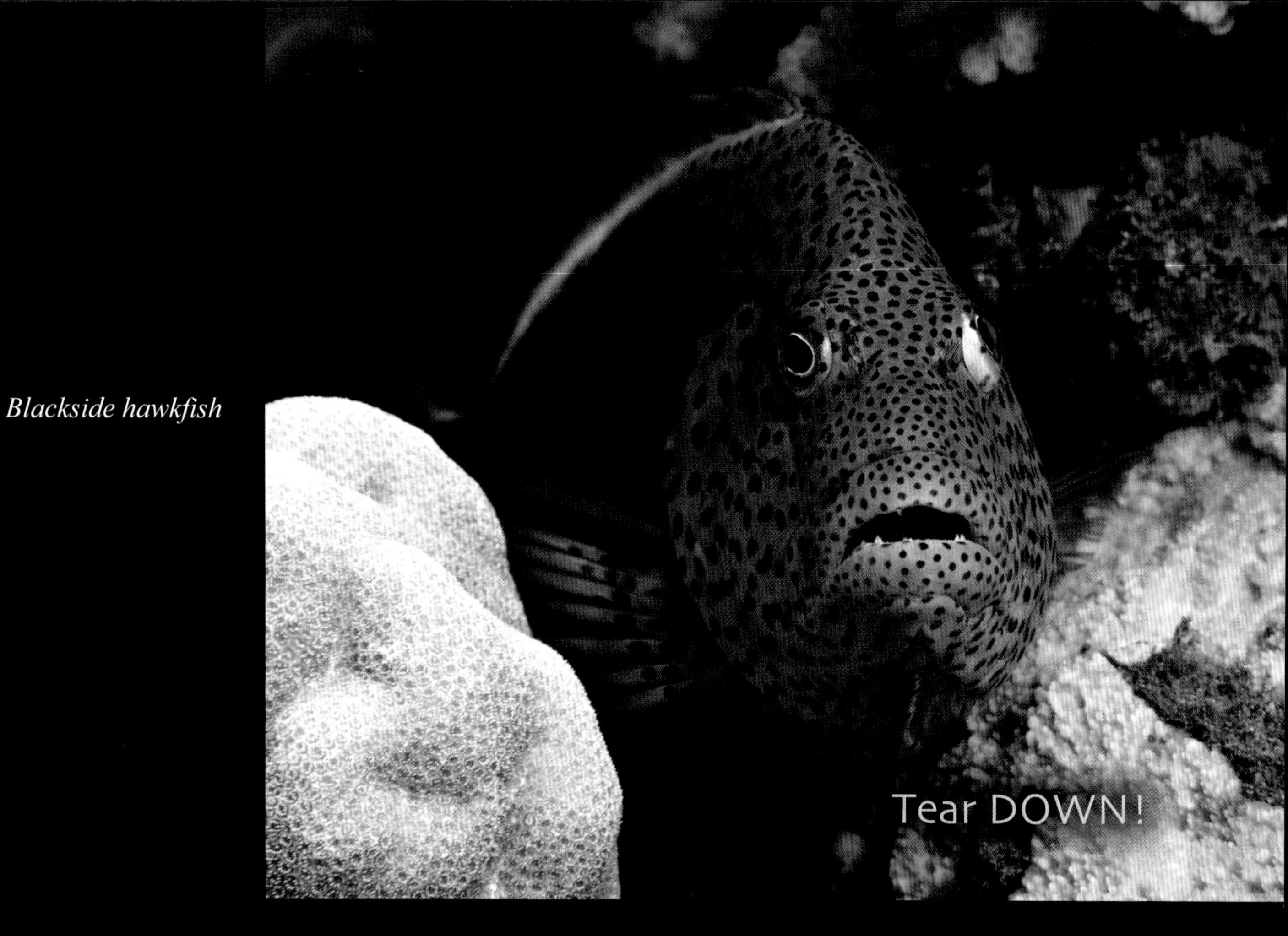

Blackside hawkfish

Hawaiian spotted tobies

You may now kiss the bride and me.

Questioning kole, or goldring surgeons

Some fish say they were in a tunnel of white light.

I don't want to live in an aquarium.

Why don't you cage your offspring and watch them?

I will pose for you, but you may not net me.

The aquarium schnorer says he sees us on nearly every reef. So why shouldn't he take us? Why shouldn't he *chob'm'n tuchus* (kiss my speckled buttox)?

Redspotted sandperch

Squirrelfish

Orangespine surgeon—an herbivore keeping algae in check.

Are you getting the message? Wake up, people, or we all through…

I'm Wanda, because Hawaiian spotted boxfish are born female, and the dominant individual morphs to male when the position opens. See the gold and purple on my face? I'll soon be Wayne, so shape up, because I ain't pussyfooting around anymore!

Barred filefish—known for social poise.

I'm Steve. I felt inferior—we couldn't afford orthodontia. Did I get confidence and a full dance card so you can mess it up?

Hawaii DLNR introduced us (bluestripe & blacktail snappers) in the 1950's as food fish. They had data, but it was worthless. Maybe they're worthless. Now we're useful as scapegoats for other invasive cultures—I mean species.

I'm a triton trumpet 40 years old. The black dot is my eye. Can you see the shame in killing me for lunch and calling it cultural heritage? Are cultures no longer responsible? Only species with consumer demand kill me for decoration. We all know what species that is.

Frankly, I'm speechless.

I'm not a vampire. I've never been to Transylvania. My real name is redtooth triggerfish, and I don't mind a pose or 2, though I do wish human traffic was not so frightening.

I'm an 8-line wrasse.
Your future may hold no-line wrasses.

I'm a disappearing wrasse, but that's meant to reference a behavior, not a crisis.

I'm a razor wrasse, and I'd like to cut the crap.

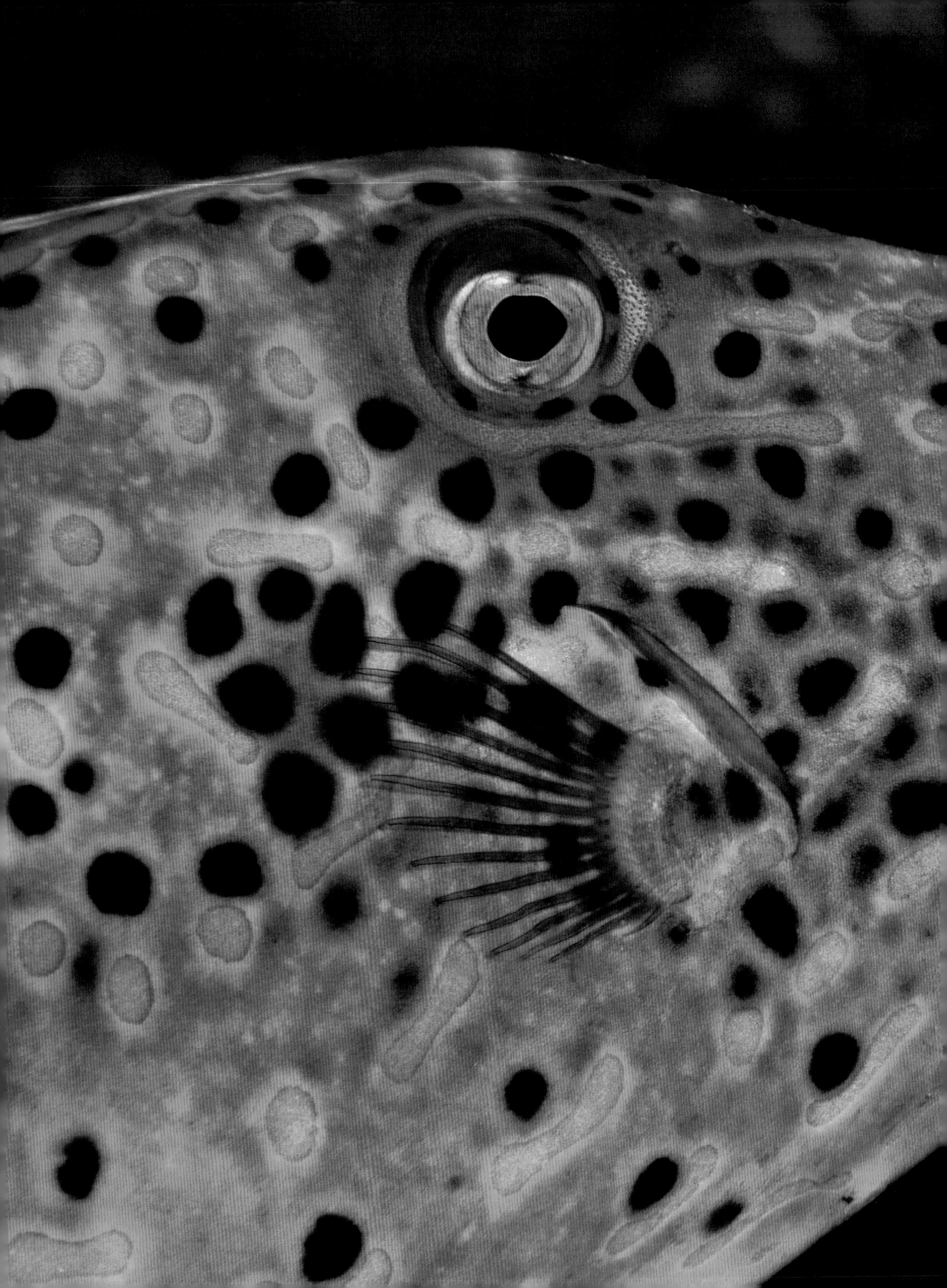

I'm just passing through. And you?

Scrawled filefish

I can’t work under these conditions.

Yellowfin surgeon, an herbivore who grazes coralline algae.

A flame wrasse female in all sincerity.

I’m not kidding, Buster!

Br'er Wrasse is a young 2-spot and a natural thespian.

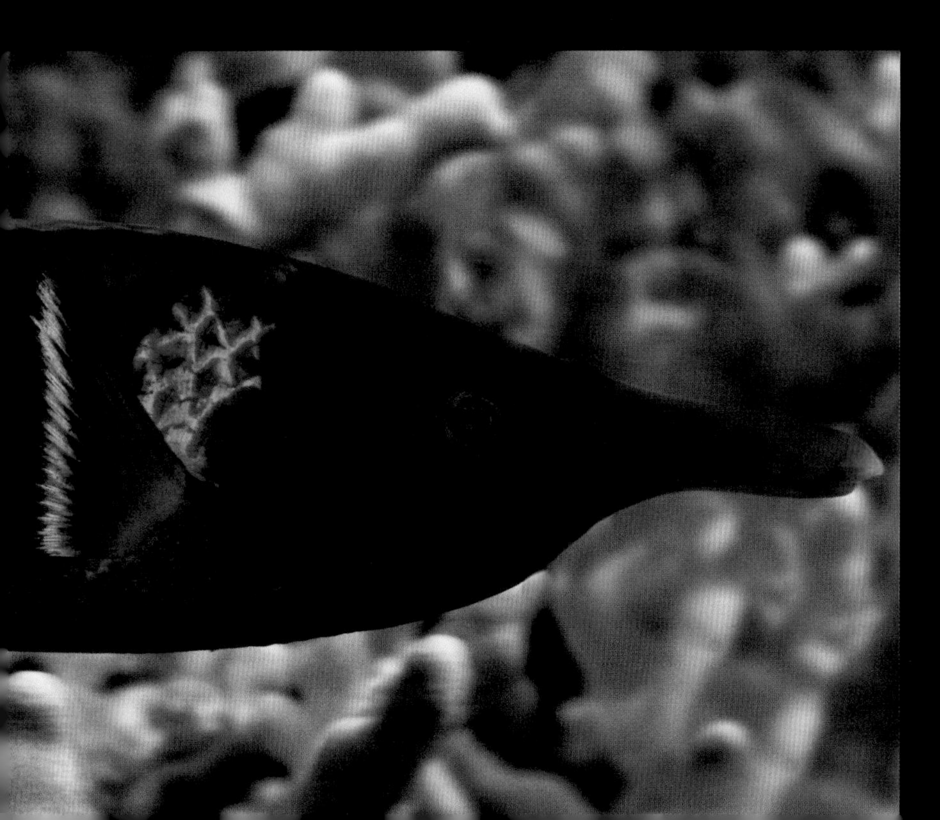

A-well-a, everybody's heard about the bird
A-well-a, bird, bird, bird, b-bird's the word
Bird, bird, bird, b-bird's the word
Well, everybody's talking about the bird
A-well-a, bird, bird, b-bird's the word
A-well-a, bird
Bbbbbbbbbbbbbbbbb, aaah…

Bird wrasse

You will not see me coming.

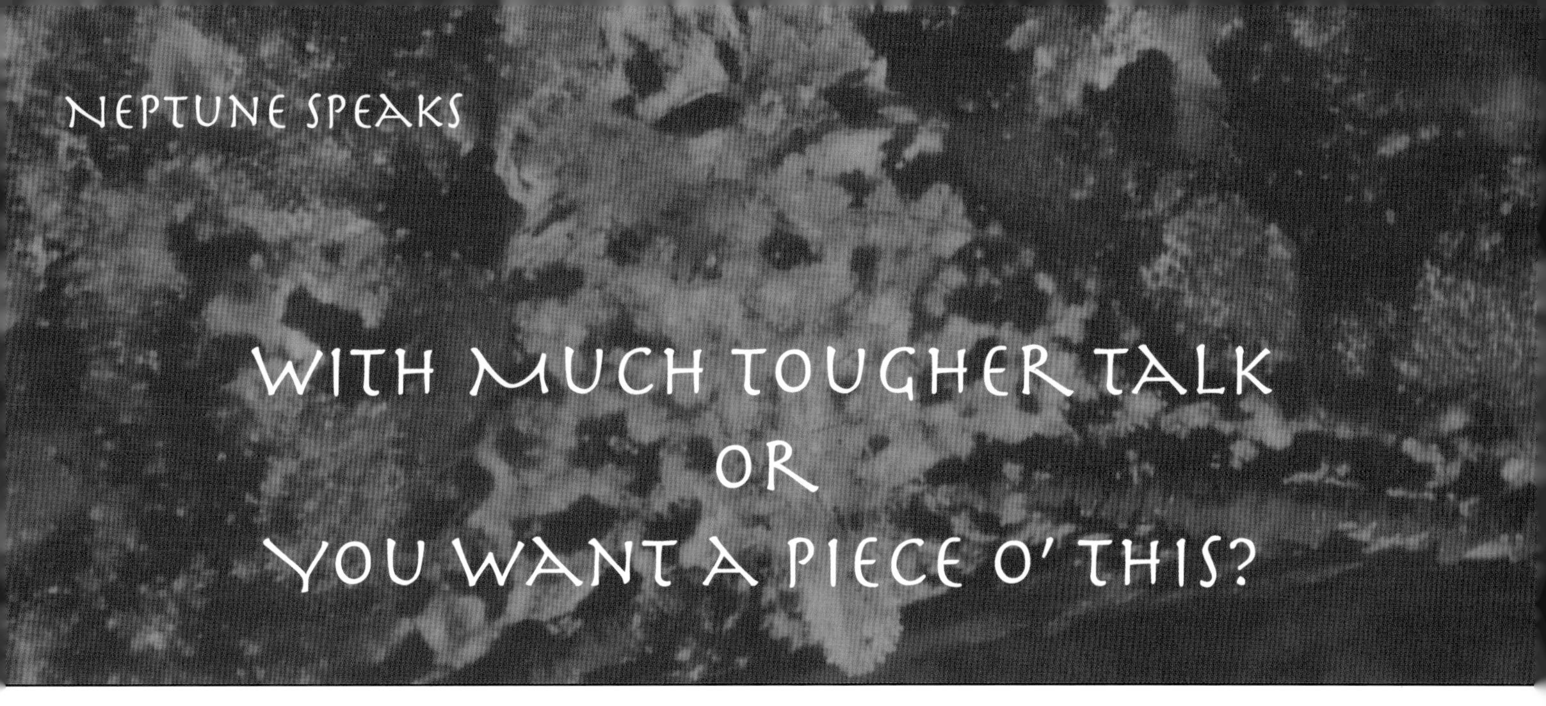

WITH MUCH TOUGHER TALK
OR
YOU WANT A PIECE O' THIS?

It's not a nice question, and tact is the better part of diplomacy. Yet some reef residents are sensitive to pressure.

Left, Commerson's frogfish. Above, camoflaged rock cod.—*Ignore at your peril.*

A cheeklined Maori wrasse doesn't give a flat flying fig for anything you might say or think. Ah, but he doesn't know what you can do.

Being as I'm a 44-milimeter arceye hawkfish, the most powerful tiny predator on the reef, and could gobble your eyebrows clean off, you've got to ask yourself one question: Do I feel lucky? Well, do ya, punk?

Axilspot hogfish

You talkina me?

What? You gotta problem widdat?

A bigeye emperor wants to know.

Anemone clownfish are as motivated as any defensive parent.

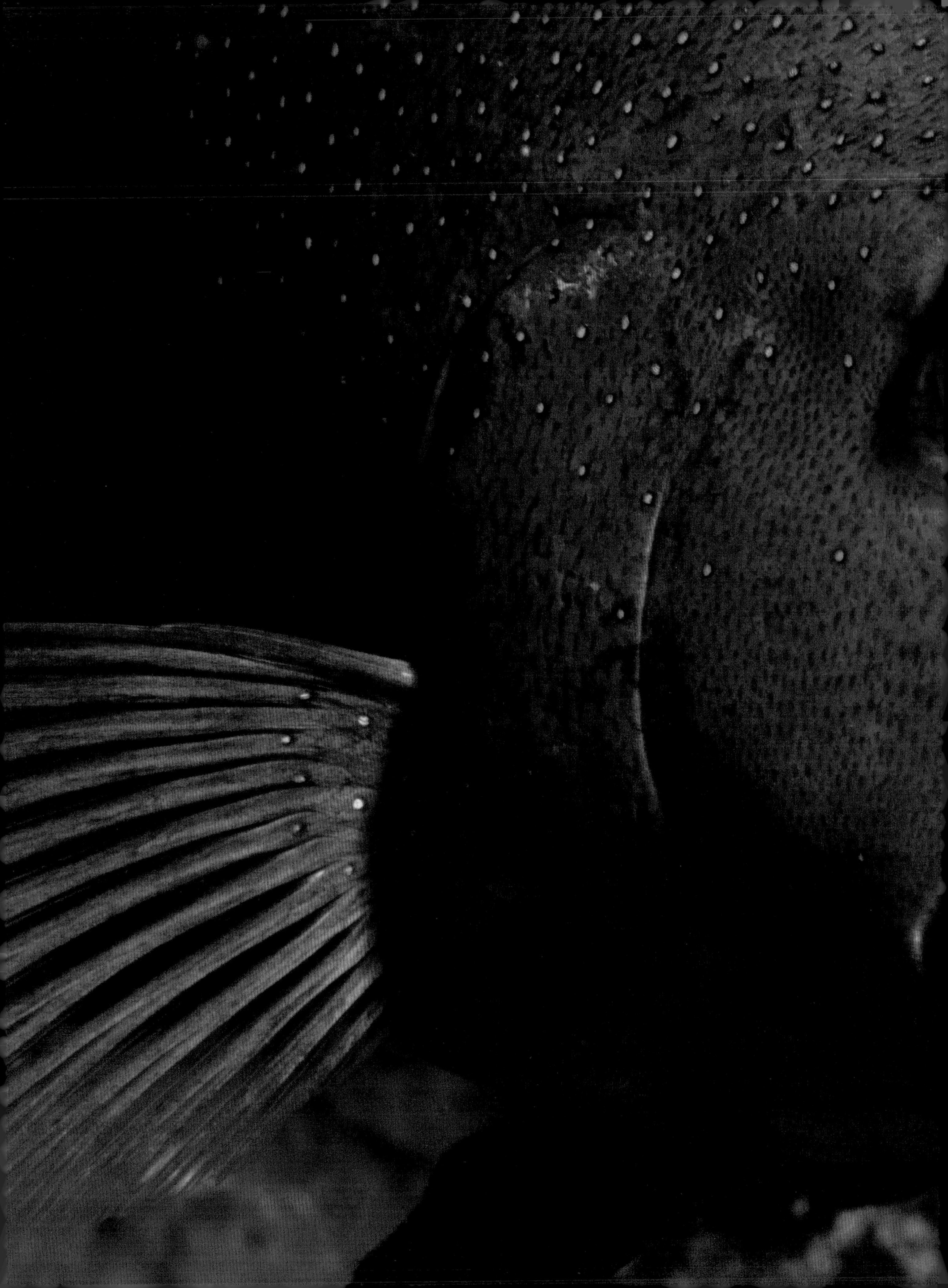

A coral grouper will face off on sheer frickin principle. These polka dots don't run.

Yellow chromis

A devil scorpionfish, above, conveys reef community outrage that so much crime against nature should make the neighborhoods unsafe. Below is a peacock grouper, or roi. I call this shot: THE DATA on DLNR. *The Hawaii Department of Land & Natural Resources introduced this "food" fish in the 1950s. It has decimated reef populations along with its sponsor, who does nothing without DATA to prove sustainability.*

I will swallow you hole.
I mean whole.*

*(Or die trying.) Commerson's frogfish

Lionfish dorsal spines are venomous but unlike nets are used only defensively. Lionfish species abound. Above may be a spotfin lion. Below is a juvenile stout moray.

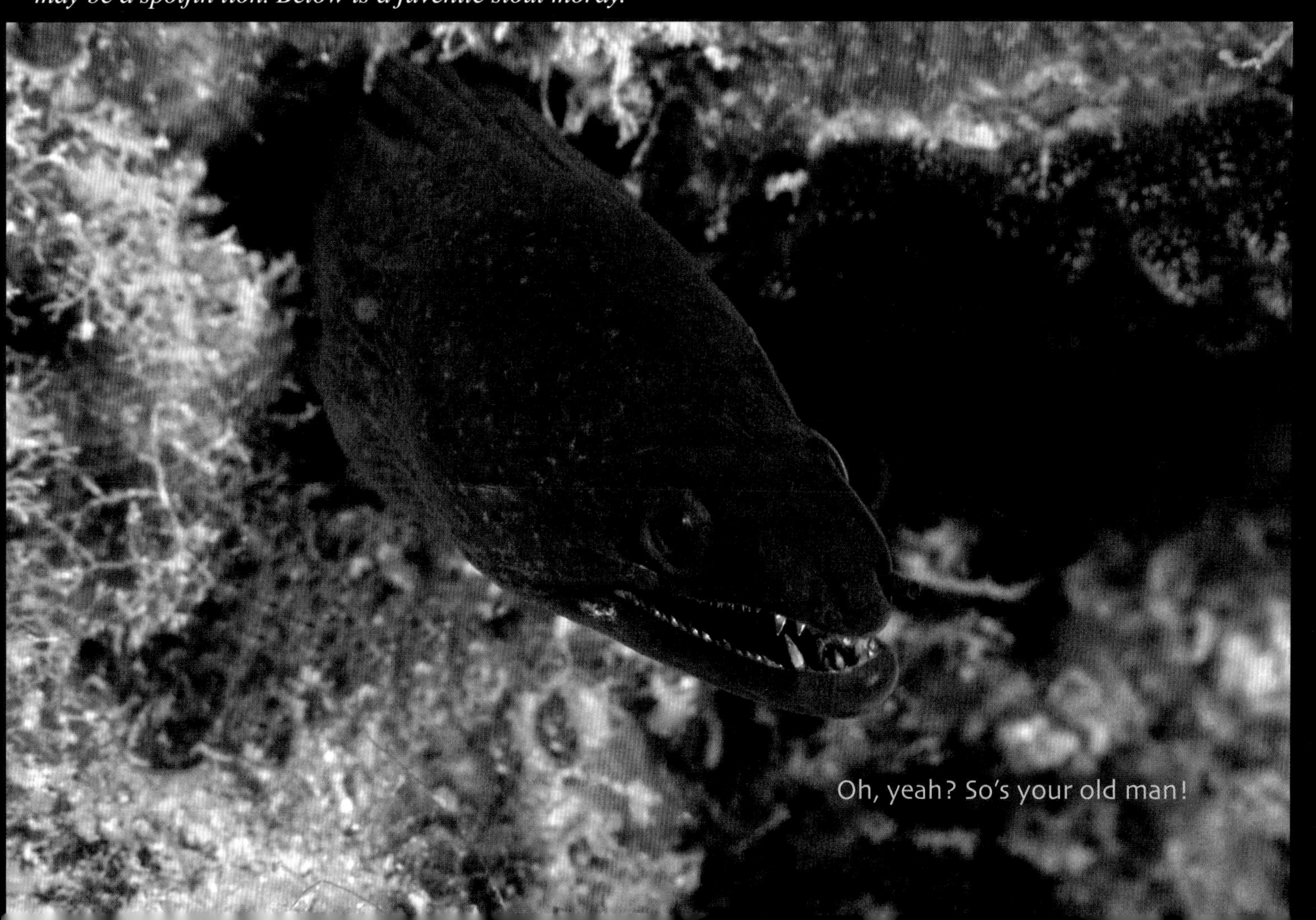

I do wish you could be snaxise for just a moment.

Lizardfish

Many humans see themselves in God's image, seeing God as an old, white man with legs, arms, long hair and a flowing beard. These attributes make God more accessible. Many humans see fish as expendable, an infinite "resource" from a limitless sea. Most humans see fish as deep-fried, wrapped in newspaper, slimy and stinky, on ice in display cases or frozen stiff or as amusements in glass tanks. Fish are often assumed inanimate, with no hair, mammary glands, lungs, shag carpet, sheetrock or drop ceilings.

"Sport" fishermen see fish as trophies much as wealthy white men saw big game animals in recent decades.

The compassion gap may stem from lack of familiarity—most humans do not visit reef communities. Other humans love fish for who they are and the unique society they provide. Kudos to those who do, for the fuller lives they enjoy.

To those who don't, Ms Titan trigger would like to have a word.

Go ahead, make my day.

All we are saying, is give reefs a chance.

Kiss me, you fool.
I think I love you.

Titan triggerfish go 2½' with a set o' choppers to put an orthodontist's kid through law school. Ouch! A nesting female will take her pound of flesh for any threat to her children. (Randall got bit on the leg, but that was likely revenge.)

Then again, on balmy bommie slopes, she could be your valentine. Now who, pray tell, is behaving in God's image?

But that's enough macho talk, fighting stance and fish emotion. I, Snorkel Bob, could use a little relaxation. A garden setting might be nice…

ON TRANQUILITY IN THE GARDEN

I'd like to be under the sea
In an octopus's garden in the shade
He'd let us in, knows where we've been
In his octopus's garden in the shade

Yellow Submarine was a cartoon for young persons of olden times who took LSD and imagined themselves underwater with wildly animated characters. And here we be again, animated and wild as ever, though these days we must venture farther afield to find the old magic—and it can turn on us quick as bad acid.

Which isn't to say you shouldn't jump over and sink 100' to the Great White Wall; it only means that beauty to the eye can be painful to the touch. Hydroids are not so problematic as hemorrhoids, unless you're careless.

Hydroids look like transplants from Pandora and may be as mystical. Their allure and delicacy can draw you in. Plentiful hydroids might make you complacent.

Philippine hydroids on a Fiji reef, left, looked arranged by Neptune, himself. How perfect, thought I, Snorkel Bob, easing in before realizing: frikkin' hydroids! A slight touch can itch and burn, but oh, the beauty is like the Tree of Life.

It's a perfect reef bouquet—but easing in again and up just a tad ...

Oops! Hovering in compliance with etiquette guidelines I was nonetheless distracted by camera clutter and the shot ahead. Sinking ¼" on the exhale for better framing, my wrist caressed the hydroid. I apologize to all careful waterdogs, to the reef and the hydroid in particular. But I mostly regret the reaction.

Hydroids can look the same in seas thousands of miles apart. These Hawaii hydroids, below, are similar and different, and come in shades of gray and black.

No big deal for an hour, till the red swelling and itching. Benedryl helped slightly, but the red bumps stayed a week and left scars. But it's not all itchy bumps in the garden. It can be tranquil, though the seemingly innocuous can still trigger a flashback...

Dandelion don't tell no lies
Dandelion will make you wise
Tell me if she laughs or cries
Blow away dandelion...

Soft corals sway in the breeze.
I mean surge.

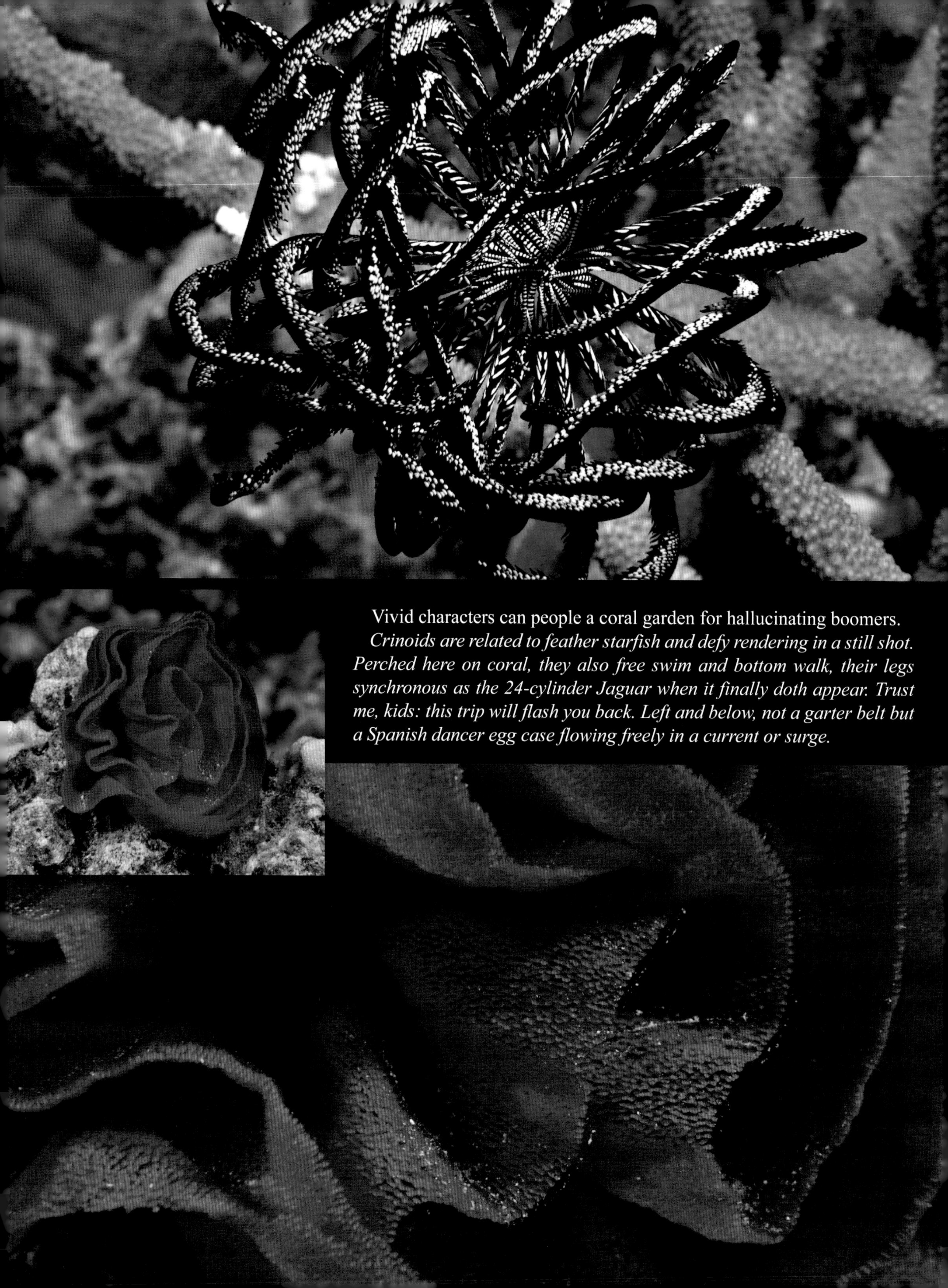

Vivid characters can people a coral garden for hallucinating boomers.

Crinoids are related to feather starfish and defy rendering in a still shot. Perched here on coral, they also free swim and bottom walk, their legs synchronous as the 24-cylinder Jaguar when it finally doth appear. Trust me, kids: this trip will flash you back. Left and below, not a garter belt but a Spanish dancer egg case flowing freely in a current or surge.

And what is a garden without slugs? I mean snails. A marlinspike auger rests in the sand. Below may be an orange gumdrop nudibranch who may grow to 1½", but this tiny pup is a spec, so it's hard to tell.

Below, from left: varicose phyllidia, kangaroo nudibranch, divided flatworm

Above, from left: Blue dragon nudi, pustulose phyllidia, Uranus anemone convention—what? Am I, SB, to know everything? Okay—genus zoanthus (Uranus).

Coral is the substrate of the reef garden, though coral is not a plant but a continually feeding animal colony that is also habitat to many.

While the blue-eyed damsel uses the coral for cover in Hawaii, the hawkfish below uses soft coral as color camouflage in Fiji—this fellow is 1" long and 100' down.

Coral takes many forms and shapes, all of which are habitat.

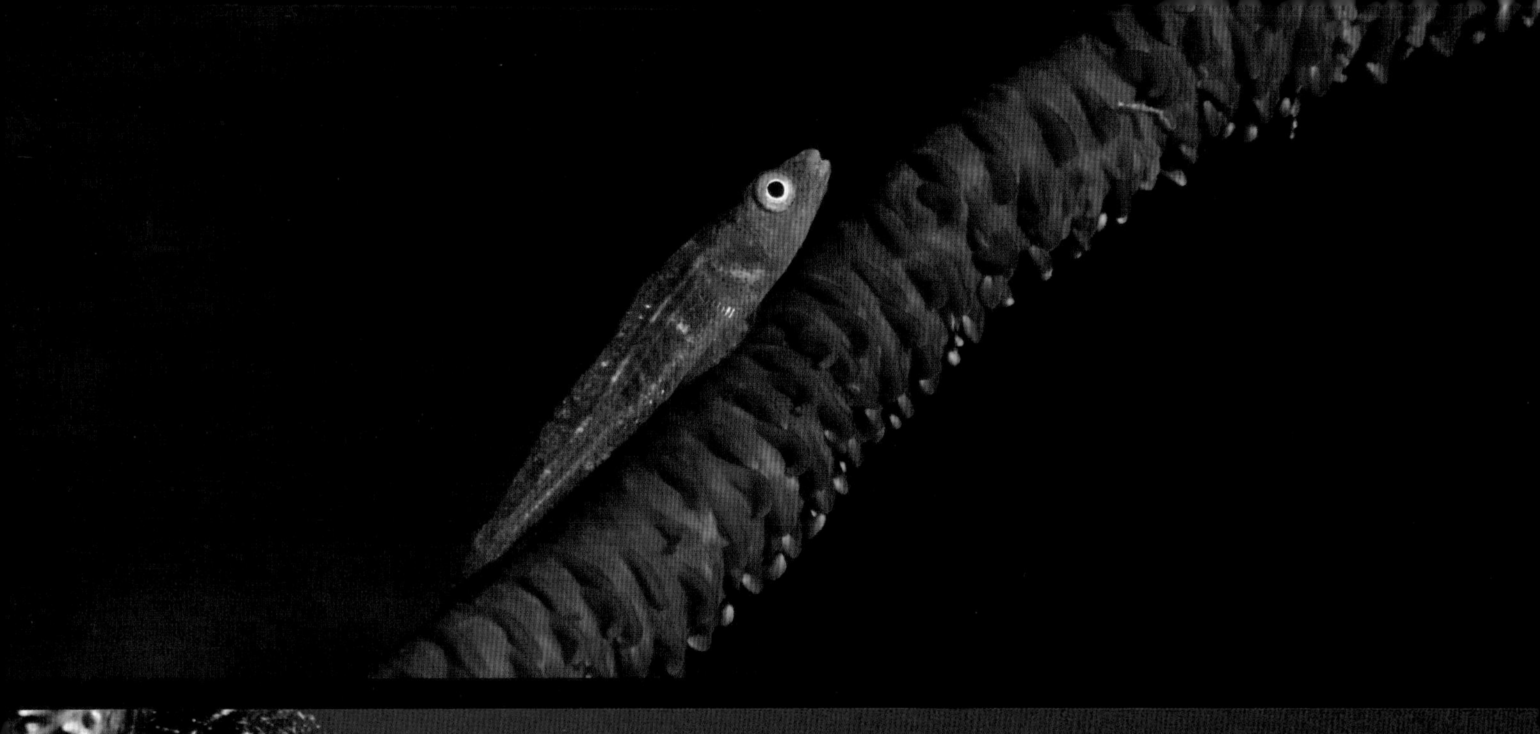

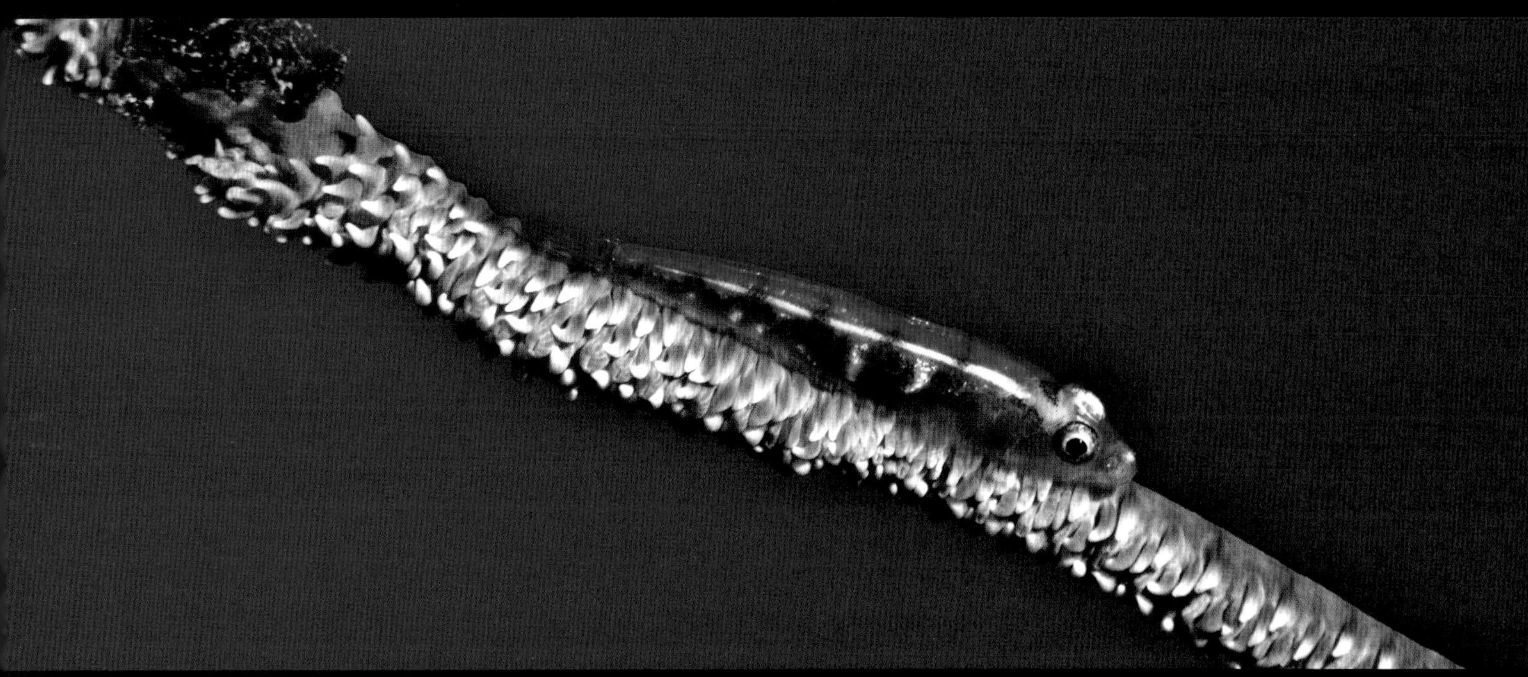

Above, wire coral gobies thrive at Mokuho 'Oniki off Molokai on red wire coral, and a few miles yonder off South Maui, too.

Left, La Vie en Orange. *Is it any wonder the anthius match the coral? This is Fiji.*

We've known for ages that longnose hawkfish live at 110' or deeper—but that was another fact based on dire circumstance. I.e., commercial divers took all the black coral to those depths, leaving no shallower habitat for the longnose hawk.

Hawaii waterdogs do not discuss exotic fish addresses for fear of aquarium scourge theft. So why would I divulge this longnose hawk's home address as 60' in black coral on a steep slope of Mokuho 'Oniki off Molokai? A fellow from Molokai said it best: "Molokai is the friendly isle, but I like tell you: we not so friendly."

Have at it, scourgers.

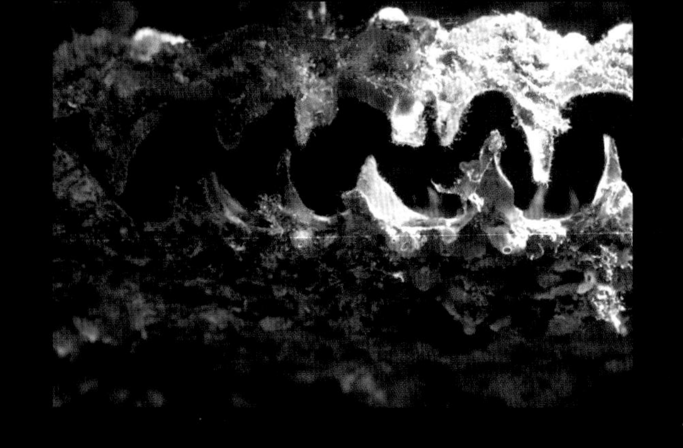

I call this garden statuary: *The Bonds of Holy Matrimony. Note the generous cavern from which a tenant may freely view the world outside. It's a black pearl oyster shell.*

Just for fun, here it is, single and free—I mean lonely—with wilderness values intact.

Garden critters are often unseen at first glance, like a leaf scorpionfish blending with substrate, encrusting corals and hydroids—or spiny brittle stars clinging like creepers.

Catch a falling starfish...

Many garden critters look like plants but are animals. A Christmas tree (top left) is a coral-boring worm who uses plumage to gather snax. The plumage comes in many colors.

Yellow crinoid (top right) in Fiji, coming on like starletfish yearning for attention.

A Hawaii tunicate is a bi-valve, who tries to mind her own business.

The coral garden rejuvenates. Waterdogs on tanks take a safety stop at 15' so the nitrogen can ease out more slowly under some pressure. We converged on a rock under the boat with Wani in the lead, counting heads as he reached for a handhold. He froze in mid-reach, sensing perilous inconvenience at hand.

Not a reef bouquet, the stonefish is extremely venomous. A stonefish dorsal injection to a human bloodstream will allow about an hour of life as we know it, without the antigen. Hardly aggressive, except when fishing, a stonefish uses its dorsal spines defensively. Still, a waterman's instinct saved the day and maybe his life. Look carefully to see the eyes and mouth—he waits for passing snax then opens wide for the big gulp.

An equally lovely note is a South Maui eelscape with color and balance of exquisite good taste.

ON A FEW OF MY FAVORITE THINGS

I, Snorkel Bob'm wary of the dog 'n pony show of the reef love phenomenon. Some genuine waterdogs are making things better and keeping things stable, but we're abuzz in platitudes, like sustainable, best management practices, "Rising Tides" and "smart" harvesting—all lip service for Mo Money. We see it in reef merchandizing too, promoting this reef or that with bombast, amazing claims and unique character. It's often hokum.

One exception is the Great White Wall in the Somosomo Straits off Taveuni Island, Fiji, where sea life is layered and prolific. Critter minions come out to see who's passing by as soft corals festoon the current to catch plankton.

Riding a current is far from a city sidewalk, and urban pressure never felt so good. Neptune's cornucopia flows forth in the Coral Triangle—that section of the Blue Planet noted for nutrient-laden current, coral and species variation, including the Philippines, Eastern Malaysia, Indonesia, Papua New Guinea and the Solomons. Fiji is on the edge.

Exotic anthius (squarespot) and many anemone lead to more…

Decades ago, I deferred to college over the Viet Nam war. I will not belabor details other than a heartland State U and vague curriculada, including a class in poetry. Oh, those were the days, my friend. I recall an undergrad inspiration, an ode in modified limerick form to an anemone from Italy who got constipated and…

An anemone from Genoa said to me
Too much spaghett could a be my epiphany…
José, can you see?
I need an anemone enema in a me.

Or some such. I got a C-, good enough to be President of the United States of America *and* secure another semester's reprieve. Who knew of the irony and greatness awaiting—that a Hoosier from the holler would grow up to be a fish-rights radical?

Healthy reefs are abundant in species variation and populations. Above is vibrant habitat for fire dartfish. Below, close on a fire dartfish:

Ambush predators are easier to frame, because they don't move. Shutter bugs poo-poo common species in deference to exotics. Neptune poo-poos such shallow values and obviously loves all species. Dwarf hawkfish in Fiji (above) demonstrate in detail, while a blackspot angelfish shows A Love Supreme.

Above left is a ring wrasse. Above right are pastel ring wrasses. These tiny fish swim synchronized till…BAM! They go limp, like dead. They drift briefly till group-swim resumes in a mesmerizing display that makes lunch seem primitive and far away.

Old adventure movies portrayed giant clams as "man eaters." Some men will poke a pickle slicer, given the chance. The great danger here stems from the most myopic species, who would take this mermaid for lunch. She's 50 and deserves better.

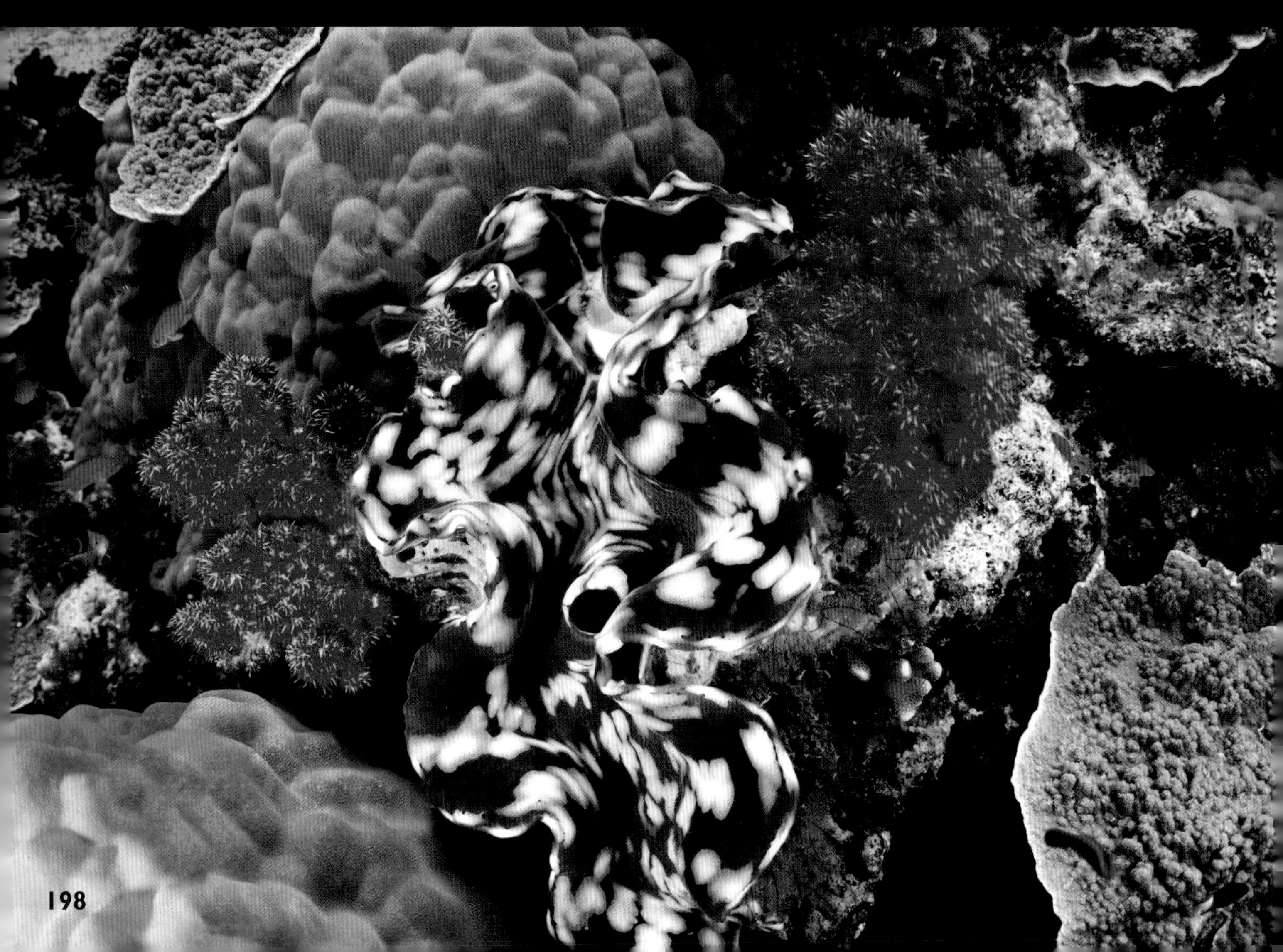

A pennant bannerfish (above) is perfectly designed for her work, while reticulated dascyllus and blue chromis (below) are the picture of reef society in casual splendor..

A princess damselfish (above) and a jeweled damsel (lower left) use color and light in their reef adaptation. A popcorn shrimp (lower right) cleans coral.

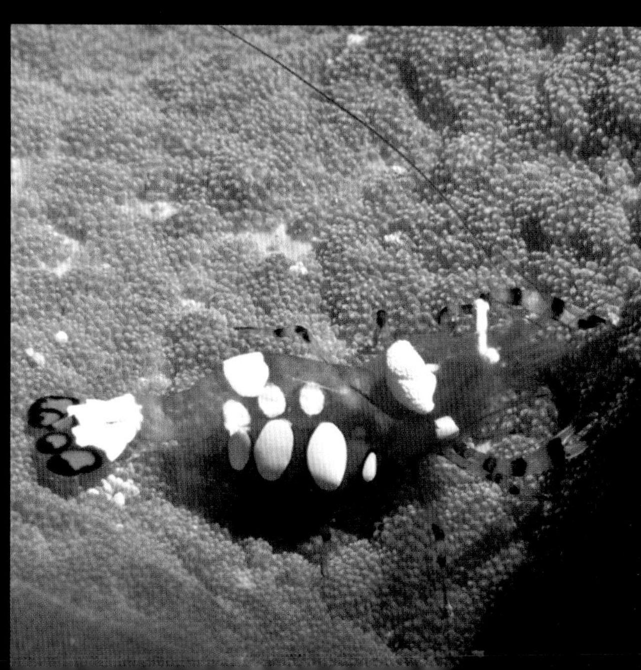

Flame wrasses are bold but shy, subdued yet flamboyant. The debs seem approachable, less challenged, receptive to a family portrait.

The beaus dart and circle, jockeying for position on the debs.

The usual challenges are: position, light, focus and angle for best advantage. But I only want a picture, and he thinks we're jousting for the debs...

A flame wrasse male is bold but circumspect.

ON NEPTUNE'S REGRETS

A grazing threadfin gives life to a sandy bottom on a cloudy day. A fish conveys trust by proceeding with personal habits in close proximity to another grazer, much bigger but with no net. This is Aloha, but alas…

A chevron tang (above) is extremely rare in Hawaii. On-line aquarium inventory sells for $150 each. A chevron is a young black tang that will turn black as an adult, though it will likely die first in a glass tank far from the home reef.

A flagtail tilefish (left) commiserates with a young trumpet (below) on the shame, shame, shame…

… as a blackside razor wrasse cannot believe her eyes and a big eye is equally amazed.

Though prayer to a personal Neptune is not often seen in the reef community, a redspotted sandperch, above, and devil scorpionfish (of all people), below, join here in group prayer. Note the frilly lappet over Scorpio's iris to camouflage the window of his hunger.

Some invertebrates are known to have more spine than some people, along with keener instinct and common sense.

Unicornfish (kala) are rebounding on Maui since the gill net ban—but the same nets are legal if set in a circle—in a failure of leadership and stewardship.

Reef issues make for strange insight. I know a fellow on Maui who netted fish all his life and demanded the "God-given right" for his children and grandchildren to net fish as well. In a meeting on the efficacy/sanity of net fishing, he presented a fisherman's rigors: "I got about 12,000 lbs akule in the net, but it getting dark already, so I decide to leave them overnight. We do that…"

Spotted Toby

He recalled working through that challenge as God-fearing Christians must do. I didn't remind him that Jews do not fear God, because my jaw was among those that dropped way back on 12,000 lbs of akule. Akule are scad. I don't know anyone who's ever seen 12,000 lbs of fish on the wing.

Big-scale soldierfish

Uncle Buzzy Agard turned conservationist after a lifetime of pounding fish. He scoffed that the younger crowd couldn't find their sterns without electronic guidance. Fishing no longer requires instinct. With electronics a few fishermen fail—but the fish always fail, because they no longer stand a chance.

Saddle wrasse

Uncle Buzzy found contrition when he found the akule motherlode. He netted the huge school for three days, till it was caught, and for years he returned to discover: no more. One fisherman can kill an ocean, one species, one spot at a time. *What's unique in the aquarium trade's unbridled harvest? Nothing.*

Barred hawkfish

NEPTUNE SPEAKS

ON PATIENCE & THE ADVENTURE AHEAD

I, Snorkel Bob, have seen actual data proving the aquarium trade "sustainable" and smoking won't cause lung cancer. I'm also told that the guests won't mind a turd in the punchbowl if it's small enough to blend with the ice cubes and fruit wedges.

Those in the lowly context (arts/humanities) suspect something amiss and often wonder: what's that hmell? I mean smell. Why, it's vested interests gathering data to support continuing extraction that's A-OK with vested sustainability.

Mimic surgeonfish, Fiji (left)

Did you say turd in the punchbowl?
Oh, Neptune, I was just IN the punchbowl!

Well, we won't be drinking punch anytime soon, and it may be a long time with so many disappointments at so many junctures, like the lamebrain TV show on Animal Abuse Planet, where 2 goombas race the clock to build a high-end death station for exotic wildlife, oblivious to all pain and suffering but their own. Or the governor of Hawaii who campaigned on his "conservation record," whose record was worse than his Republican predecessor in his first year of office—including appointment of an aquarium collector as Director of Land & Natural Resources, then stonewalling all polite efforts to encourage him to BE a conservation governor. Lulu knew.

The reef liberation campaign continues as it must, as a labor of love. Martin Luther King knew he might not reach the Promised Land, as Moses knew before him. Yet they plodded on, compelled to choose the good. Well, it's not so grim out here on a horse with no name. A campsite feels homey, and we have our high times.

Like when Maui County turned the screws on the aquarium trade, driving the biggest collector and his shrill wife/dealer out of town.

Or when Hawaii Island (the Big Island) County Council voted to ban the aquarium trade statewide!

And so did Kauai County Council!

We're far from ending the dark ages of ocean exploitation. Wherever humanity approaches the sea, the sea will struggle. Human propagation beyond the bounds of good sense will compromise any wilderness and threaten to make it a wasteland. Which brings us full circle, back to humanity that remains blind to the difference between wilderness and wasteland; one is all natural, while the other is all used up.

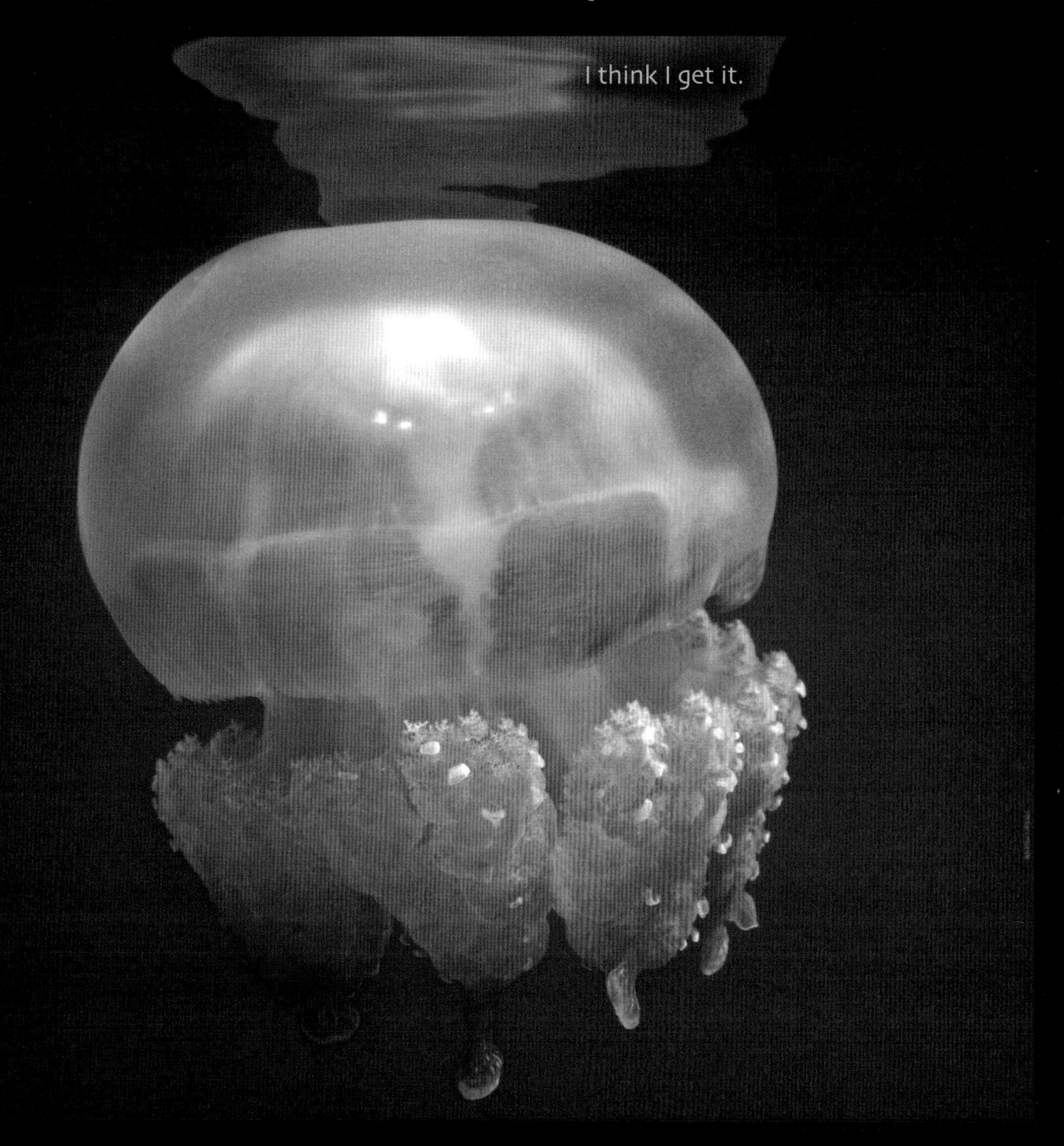

Which isn't to suggest that humanity should emulate the jellies, from whence it came, or that we must ponder objective, consequence and practicality to painful extreme—it's like the guy in the window seat said right after screaming, "There's a man on the wing!" He didn't even pause for a heartbeat before the follow-up, which seems obvious, derivative and obsequious: "We're all gonna die!"

Reef critters remind us that a good life is a series of small improvements with regular breaks for art, love and fun. Whether by choice or internal combustion, we celebrate, and therefore we am.

And that, Ladies & Gentlemen, Boys & Girls, brings us to Neptune's communion with a segment of humanity that has chosen the glorious over the nefarious, that has shown wilderness wisdom in the world peace now available in…

I too am very delighted!

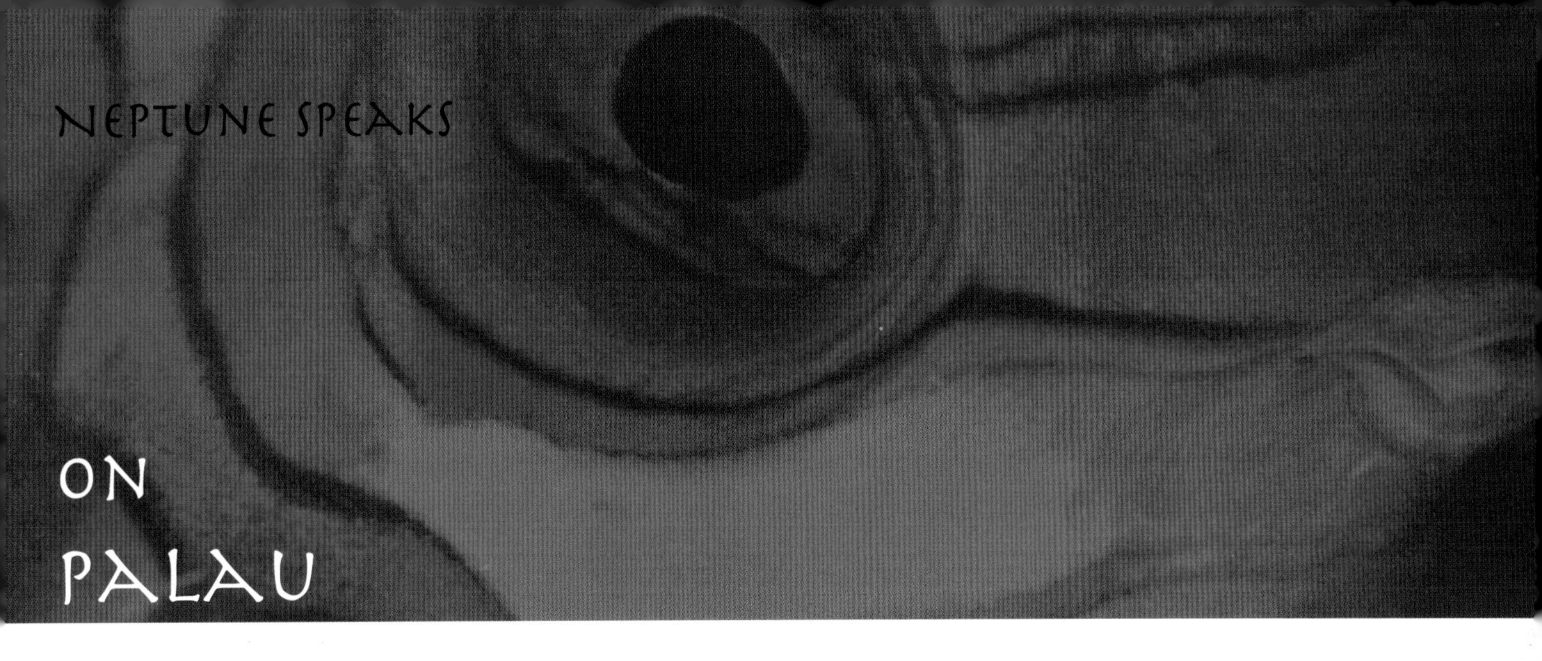

EXPORT OF LIVE REEF FISH
FROM PALAU
BANNED!

May 11, 2008

I am very delighted to report that Palau's Legislative and Executive Branches have come together to support individual states in bringing an end to the controversial practice of "Live Reef Fisheries" in Palau with the recent signing into law of a bill outlawing the destructive practice. The people and leaders of Palau's northerly states of Ngarchelong and Kayangel sought the assistance of the national government to help protect Palau's environmentally important Northern Reefs.

The issue of Live Reef Fisheries first came to public attention through postings made on this blog dating back to 11/19/07 and continuing through 12/08/07. After the story on Live Reef Fisheries first broke on this blog a great deal of public debate and very close scrutiny took place on the environmental impact caused by this unsustainable and destructive fishing practice. States then appealed to the national government for assistance in ending the practice.

President Remengesau, Senator Alan Seid, Members of the OEK (Congress), and the leadership of Ngarchelong and Kayangel States are to be commended for their actions to protect Palau's natural resources and marine environment from unsustainable exploitation.

—from the Sam's Tours blogsite
Sam's is an excellent dive/snorkel operation in Korror, Palau

Now THIS is a celebration! And with no further adieu we will cut to the chase—no, not the chase, because Mandarins, like all dragonets, are small, shy and reclusive, so a chase would only discourage engagement. Better to hover, neutrally buoyant, breathing easy as an old dead log, some of whom may still breathe, if you squint and open your mind like a young, idealistic Mandarin fish.

I, Snorkel Bob, avoid disclosure of home location on any exotic fish, and photo tech is also personal. But guidelines can have exceptions, and here we are. Mandarins are notable for dramatic coloration, lust and circumspection, which add up sensibly if you think about it. That is, they conjoin daily in a set interlude of about 20 minutes at dusk, coming away from shelter into the water column and predatory exposure for more elbowroom in which to freely embrace in the gillbreather version of boom shacka lacka. Timing is practical. Dusk is very low light, minimizing visibility for predators, and dusk is also a shift change, when diurnal critters are tucking in for the night, and the nocturnals have yet to arouse. Why the amazing coloration in light of predatory concern? Obviously, to make the mating game more alluring, dashing, romantic and flamboyant, even in low light.

Favored habitat is rocky, with pukas and crevasses, like walls. Sam's Tours is a great dive/snorkel outfit in Korror, with a dock out front. Despite excellence, Sam's is remiss on bottom hygiene, and I, Snorkel Bob, drifted among scads o' fuzzy fishing line, old bones, plastic crap and trash, thinking that this is not… Wait! There!

Love at first sight was shocking. I struck stillness. Tiny Mandarin vanished.

I lay there a long time and longer still, till Tiny peeked out. He hid again for a long time and longer still. He peeked again and darted out for a trial face off—then darted back in. Then he ambled out casual as a carefree galoot at happy hour, munching the schmutz like he and I were snorkel buddies from way back.

Sam's wall faces north, making dusky shade nearly all day. It's not wilderness—tourons take pictures more regular than tidal movement, and the critters get familiar. Which is not to discount the difficulty of a family portrait on such a tiny, skittish fish.

A teleconverter is a lens attachment for more detail—for every dimple and chin whisker a little Mandarin is willing to share. But in low light? And hazy water? You could get skunked! And then what? Well, it's a crapshoot, and there's always tomorrow, except that a loss far from home cannot be covered, only deferred. But never mind the negatory; this is a celebration after all, even with focal distance so achingly shortened as to render any behavior forward, if not obtrusive. But then…

It came to pass, that the little fellow was safely in the can, on his way to immortality while finding comfort at home in the company of a huge, cumbersome marine mammal, for whom he came forth with flattery and the ultimate macho praise…

Romance has always been a big part of showbiz. Why would anyone want camera time, if not to attract diddle mates? The great milestone here, however, was not spawning potential but photo-ops with other denizens of Sam's wall in maximum magnification. Like these network pipefish…

The wall is made of giant concrete blocks encrusted with coral and fuzzed with algae. Between the rows is a narrow crevasse called home for this ringed pipefish.

Again, we can only speculate that habitat so frequently visited by human tourists with cameras tends to desensitize normally shy residents, who continue with late lunch even as an interloper horns in for the close shot.

Above, network pipefish. Below, tiny patootie.

Hardly surprising to an inveterate snorkeler, the 5-20' zone is a rich community and great place to linger. An idle recollection emerged of my 8th grade teacher, Mrs. Williams, who was old even then and has met her maker by now, or she's still oppressing youth at 120. Oh, some teachers are bad; she seemed so keen on tormenting the little Jewboy with the nerve to be in her class. In the spelling bee final it got down to another kid and me, and I got the winning word, *appropriate*. How ironic, I thought, even. She pronounced it appropiate, so I repeated: "Appropiate?" She repeated too: appropiate. So I spelled appropiate.

And she said, "NO! You left out the r!" The other kid won. I'm certain nobody else on the face of the earth remembers that moment. Then came Easter services across the road from school, in a church. This was Southern Indiana, long, long ago. I told Mrs. Williams that Jews don't go to church. She gave me a mark in (bad) citizenship. In the church, I told Mrs. Williams that Jews don't kneel to giant crucifixes. She gave me another mark in (bad) citizenship. I walked out of the church for my 3rd mark in (bad) citizenship, and 3 marks meant failure of the 8th grade. At home Old Mom said, "Oh, you'll fail. Over my dead body you'll fail." That was the year the Supreme Court of the United States of America (Earl Warren) confirmed separation of church and state, giving little Jews, Druids, Visigoths and Others federal protection. Neo-fundamentalists across the nation are STILL crying foul. I got a gang of my little 8th grade pals to join me in a grand TP of Mrs. Williams' house soon after Easter services. She watched out the window, and the next week we got called into Mr. Rohr's office for Christian mercy as dispensed by Mrs. Williams, who would decline pressing charges. I smiled, verging on delinquency even then, asking if she'd like to decline the Anti-Defamation League too. She and Mr. Rohr, ignernt of leagues and such, stared like the voice of you-know-whom spewed from my mouth hole.

And there she is, after all those decades, living peaceably along the wall under Sam's dock, reincarnated as a marine asshole…I know: I should have asked her to use the word in a sentence. Maybe someday I'll join her, filtering freely in the flow, a crusty but serene barnacle under Sam's.

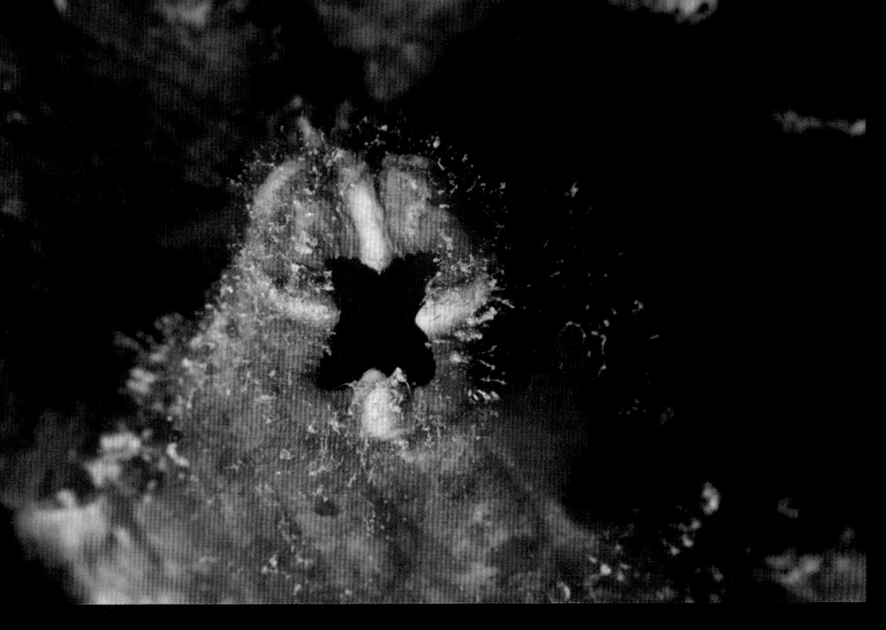

Mrs. Williams in the next life, a humble squirt under Sam's dock—and to remind her of what she barely missed, a devilishly prickly starfish below.

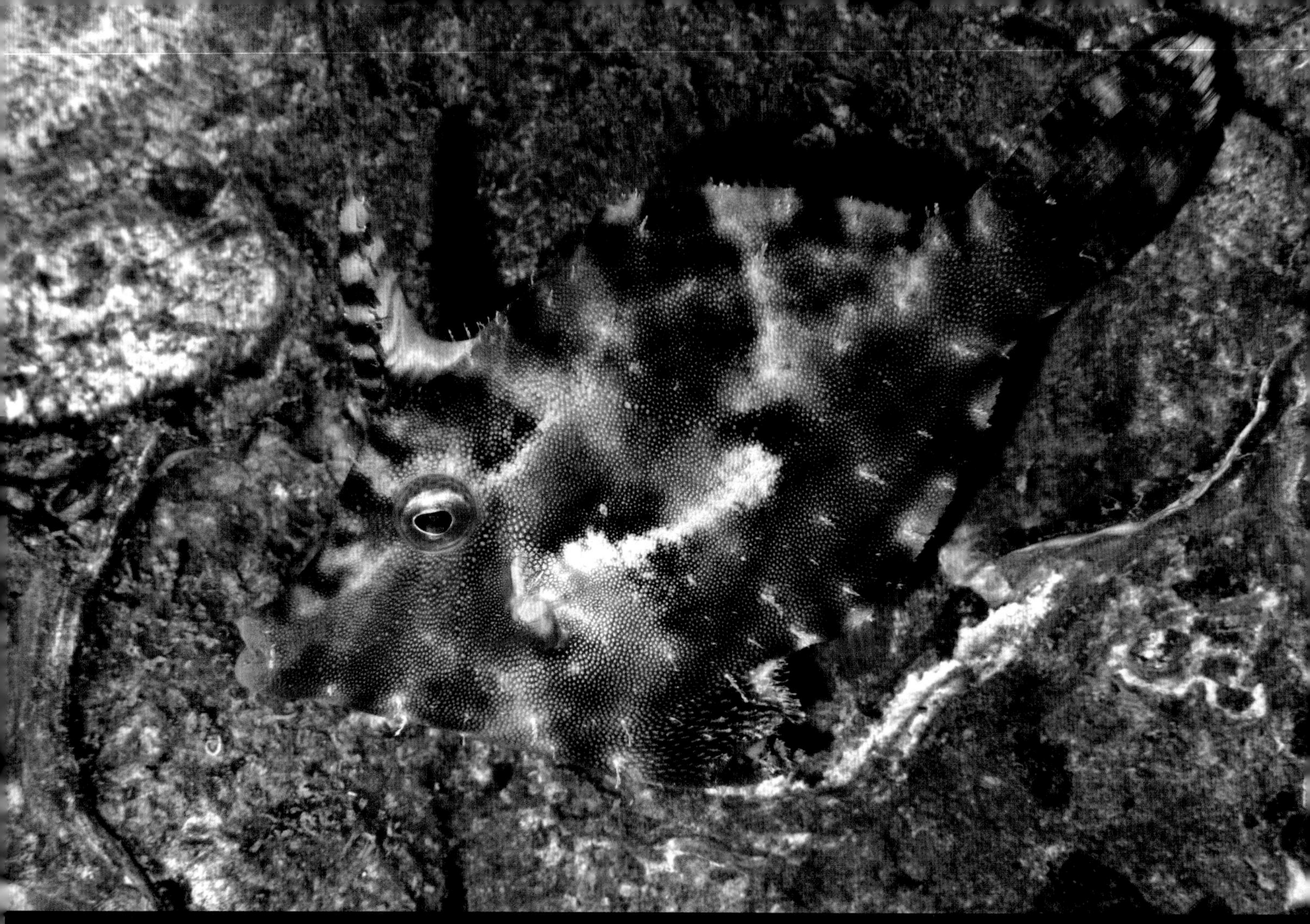

Meanwhile, back in the moment, life on the wall goes in and out of focus, color and splotch, convoluting till reality merges with illusion. And out comes life itself as a *whitebar filefish* blending perfectly, till he grazes on a fishbone cast aside by a human.

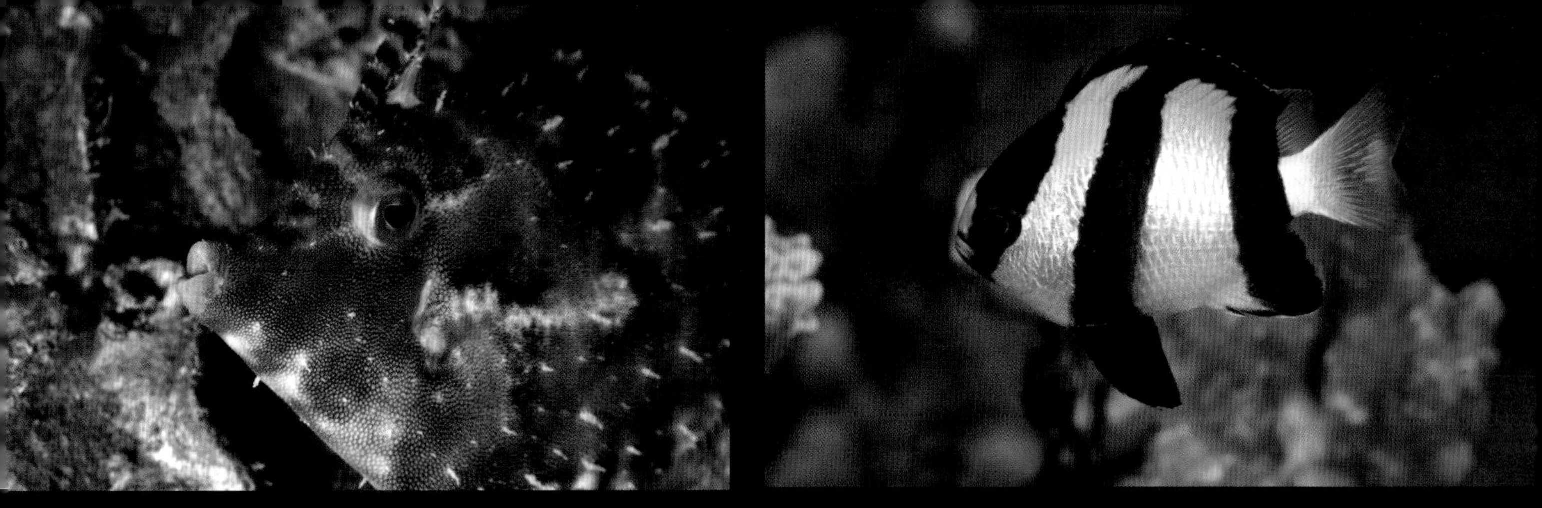

Another tiny patootie, above right, this one a humbug dascyllus.

Meanwhile, at the far end, a pajama cardinalfish, waits with dazzling color *and* a floppy 2nd dorsal.

A shorthead fang blenny is busy as the next fish making a home, but takes time for a family portrait with aloha.

Darting out with a nip (no tuck) is a little jerk, daring, challenging and scaring the snot outa the hunky marine mammal with the camera, the obstreperous jerkfish. He's unsubtle and maybe a bit pushy, but there's really no rush.

It's time for you to go.

Obstreperous got it right. You should be going, really.

Okay, I'm going. This place is done anyway, with nobody left to shoot… Wait!

It's a crocodile fish who looks more like an alligator than a crocodile but should really be called a teddybear fish. Check the shayna punim on this pup—and the frilly lappets over his irises for better camouflage… Wait!

Oh, Neptune, *a sea snake*!

The banded sea krait keeps a gentle disposition in these waters, unless provoked. Sea kraits are indeed snakes, reptilian lung breathers who surface for air. Flat tails provide propulsion with no ripples. OF NOTE is their venom; though fatal to humans, it emits from fangs back in the molar section. The krait needs loose skin to gum back to the fangs for invenomation—hey, wait! It looks like a school of loose skin folds…

It's shrimpfish drifting vertically, heads down. OF NOTE: Anita is a world-class snorkeler who goes overboard in a blink to view seamount summits in the open ocean. Unburdened by shrinking camera syndrome, she carries a point'n shoot that cost fractional pfennigs on the dollar and captures many turtles, sharks, parrots, frogfish, clowns and the rest. She cares nary a fig for backscatter or compelling detail and remains happy, as you may too, unless you're compelled to get it right—wait. I didn't mean right as opposed to wrong. I mean…you know what I mean. Don't you? And now it really is time to go, to seek what we shall find, to cut, as it were, to the chase…

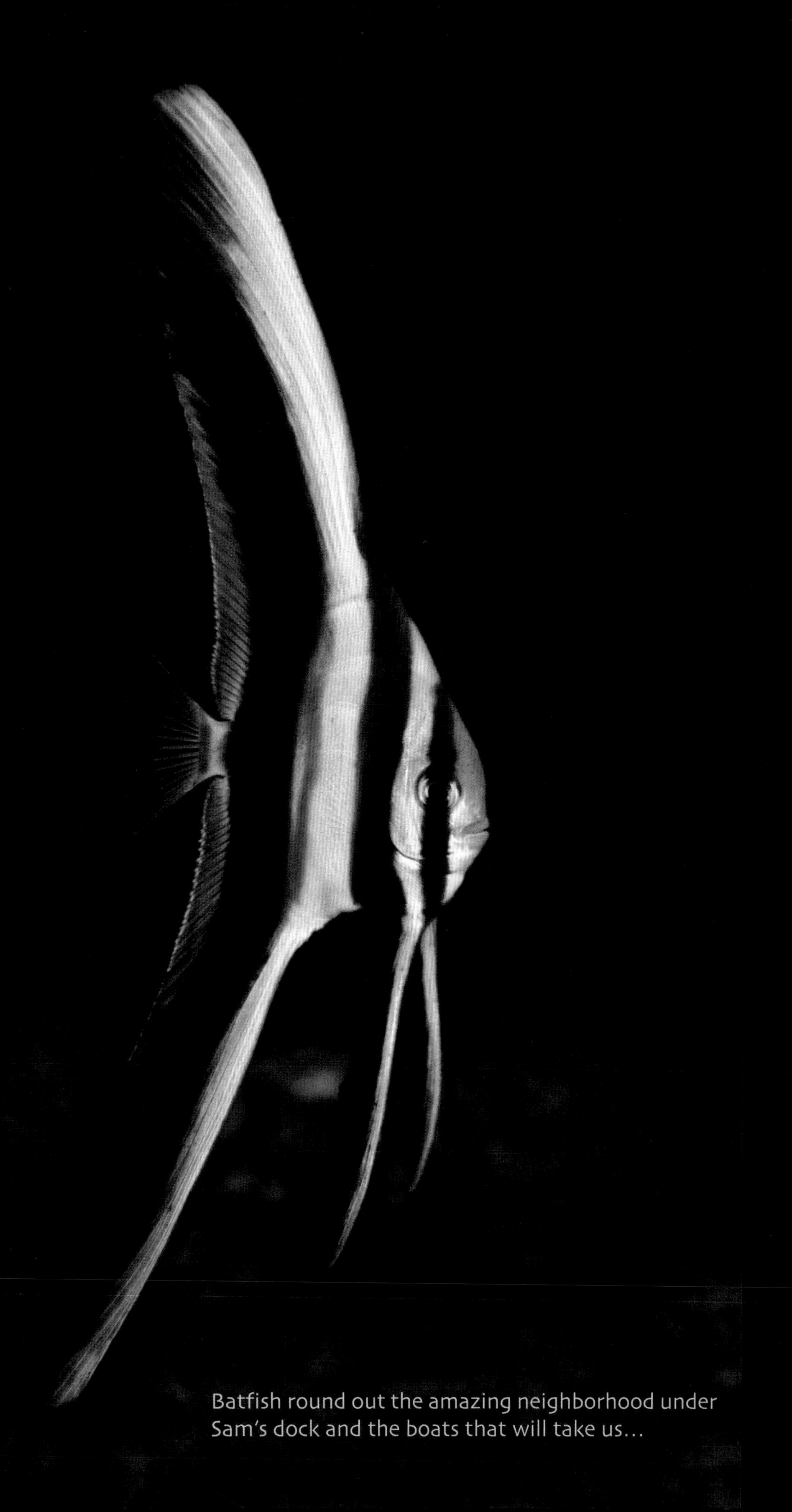

Batfish round out the amazing neighborhood under
Sam's dock and the boats that will take us…

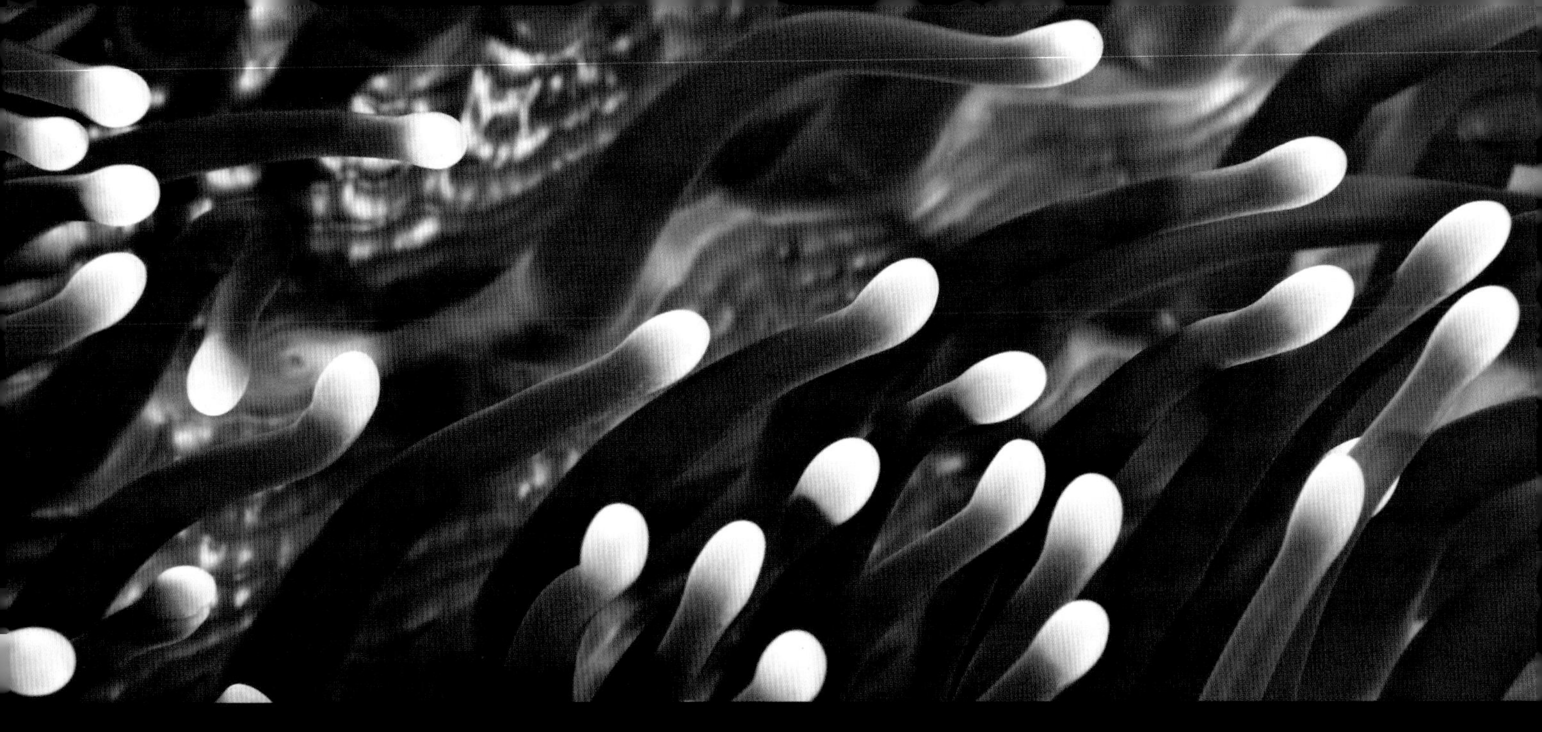

yonder…

Some divers view exotic sites as groupies see rock stars; *I really need that one*. Merchandizing is not a wilderness value. Exotic sites may be waaay the hell out there—with heavy traffic. Most dive operators visit these sites and hook in. I.e. a big fishhook with the point ground to a nub and the barb removed is tied to a hogclip by 10' of braided nylon line. The hogclip clips onto your buoyancy compensator. Hooking in to substrate holds you in the current. Reef hooks damage reefs. Current is part of a greater flow, meant to be ridden, not jammed. I, Snorkel Bob, don't like reef hooks.

Hokay, Blue Hole is literally on a 20' bottom that felt perfect for a guy with a camera, rife with exotic critters, like a black lionfish, featherdusters…

Black lionfish (above), featherduster (left)

NOTE: The aquarium trade devastated Kane'ohe Bay on Oahu, removing featherdusters from their habitat with crowbars. When I, Snorkel Bob, pointed this out, they cried, "Liar! Liar! We never used crowbars!" Oops, they used tire irons. Pardon frikkin' me.

Who dat?

The shallow ledge is so diverse, exotic and tempting…

Yellowmask angelfish

…that I, Snorkel Bob, looked up after barely a minute to find everybody else gone down the rabbit hole—I mean the Blue Hole.

Better scoot, sucker.

Blue Hole drops another 40' past micro-neighborhoods festooning the walls with encrusting corals, soft corals, ascidians, hydroids and sponges that go beyond 100% coral cover to 2, 3 and 400% in layers of life. And critters…

Did I, Snorkel Bob, ever tell you about the magenta dottyback who fell in love with a starfish? Turns out the starfish used to be an anemone, but it didn't feel right, so he saved his money for the surgical procedure, but then he met dottyback, and they shacked up right away and…wait… Where is everybody?

Magenta dottyback

Down, down, down to about 60', Blue Hole opens into a huge chamber with a bottom about 40 or 50' below that. Here too the walls and ledges are thick as Manhattan with almost as much predatory action but hardly the pollution. One side of the chamber is an arch, floor to ceiling, and beyond the arch: Blue Wall.

Drifting on a ½-knot current makes perfect sense. Just as a moonwalker gets 20' on a casual jump, so a drifter can fin once and glide with the fishy minions. Among the dazzlers of the deep: anthius in purple, purple queen, amethyst and squarespot.

Above, purple queen anthius, male in front on Blue Wall. Below, purple anthius, female left.

Above, purple queen anthius

Below and facing page, Squarespot anthius

You cannot be here without sensing Neptune watching.

Pennant bannerfish (above), sweetlips with sweetlips pup (below).

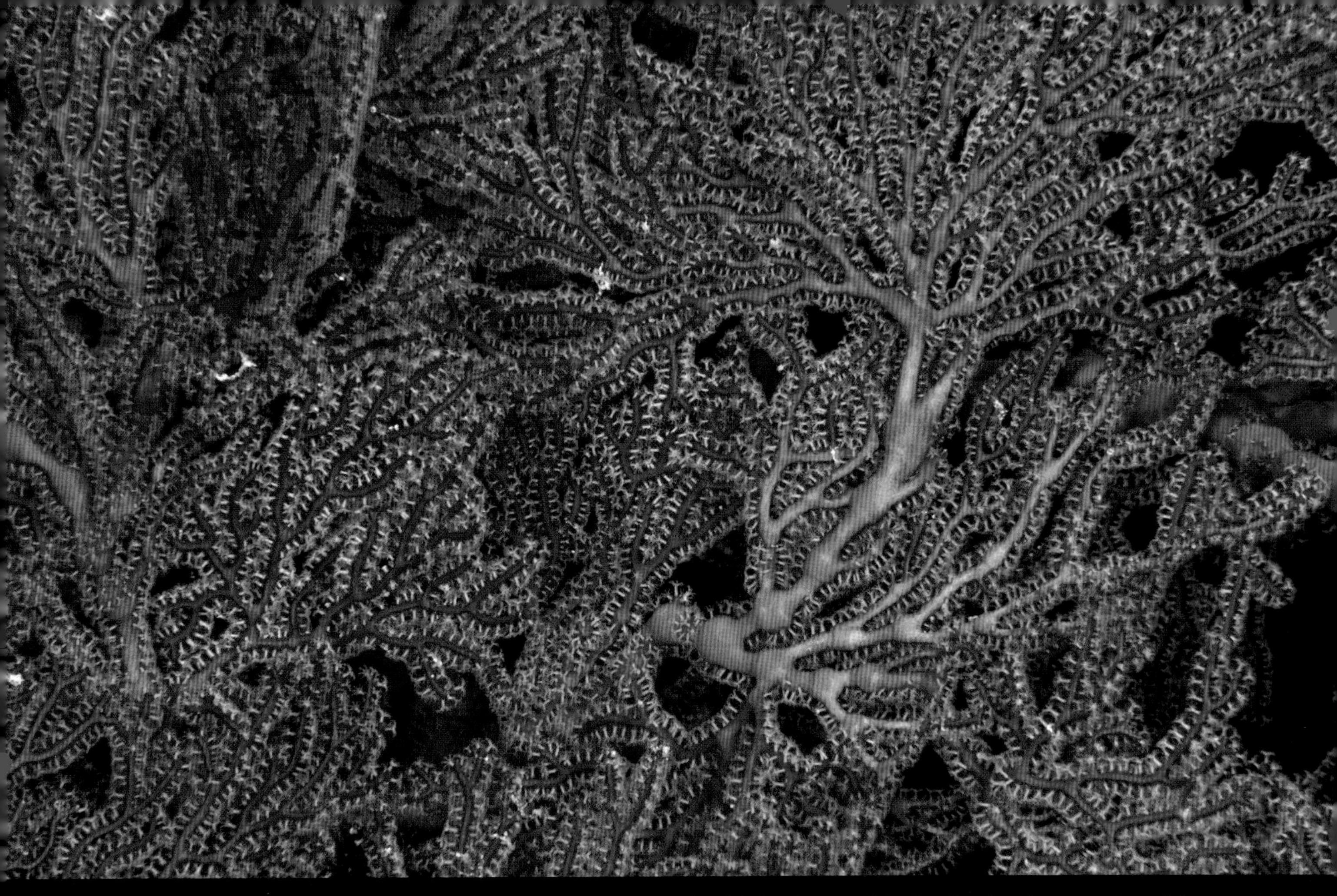

Every polyp is a critter on the sea fan above. Below, a crinoid, in the starfish family, holds onto the whip coral with prehensile "feet" and swims freely at night.

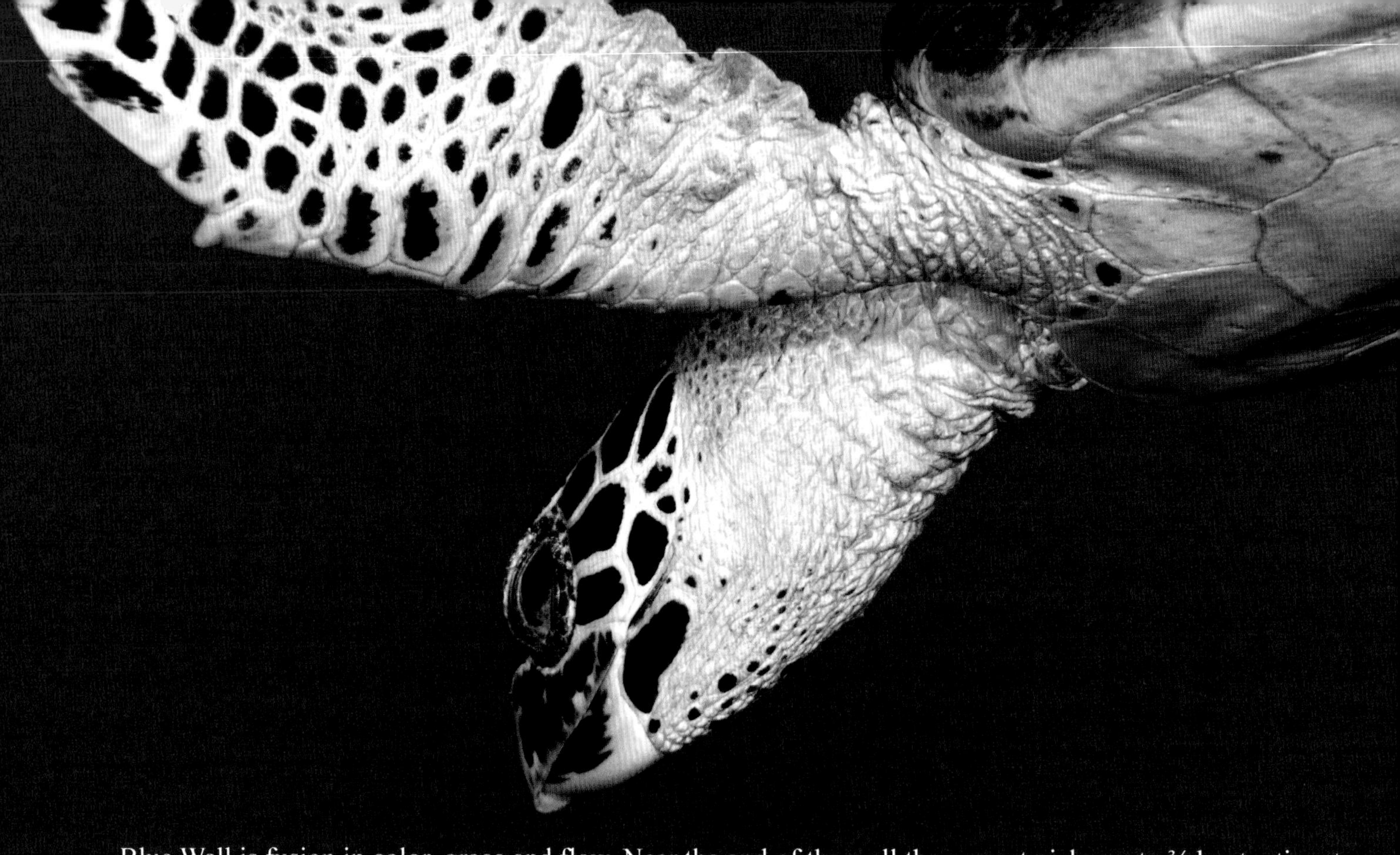

Blue Wall is fusion in color, grace and flow. Near the end of the wall the current picks up to ¾ knot—time to move out, to avoid the cross current over the top that could pin a person onto the wall for yet another layer of life, a flailing one. Moving out is easy enough, except for the endless distraction, like an old friend overhead, pausing with Aloha, a hawksbill turtle, a survivor. Hawksbills are endangered everywhere. Heller's barracuda are smaller than great barracuda. They school, often surrounding a reef dog for the family portrait.

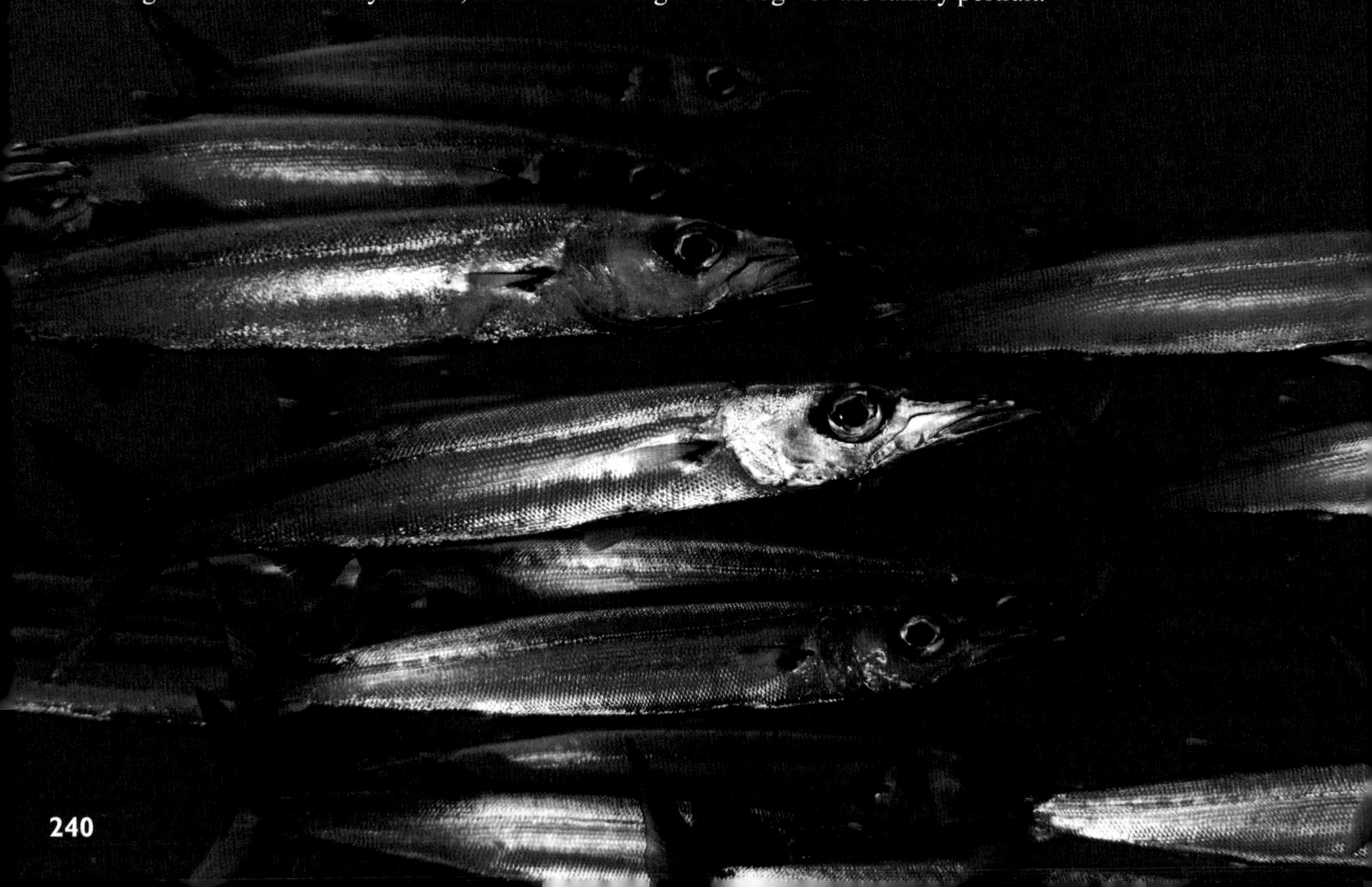

Fire dartfish beckon from the wall...
As we move to the depths, where gamefish cruise. If we stay calm, our friends in gray suits won't scurry off.

This is not complicated but warrants undivided attention, even in view of such splendid distraction, like African pompano cruising past at the perfect age, in which their threadfins have not yet faded into adulthood.

Current jolting to 2 knots feels urgent, slightly faster than the Olympic wunderkind. Then 2½, who'll give me 3? Is that 4 knots? Okay, the drill is to move out from the wall and ascend to 30' to pick up the freight train highballing over the top and onto the flats. Then stick your camera up your nose while you fiddle futz with the hogclip on your D-rings to release the giant fish hook stashed inside your BC, then hook on and…enjoy the jam. Meanwhile, looking aside will remove the mask, complicating safe ascent and the otherwise lovely aspects heretofore.

3 forces converge: 1) 465 lbs of air remaining; 2) 2 minutes till a decompression stop; 3) quick application of a rarely used college degree. An instinctive dislike of the sound, feel and concept of reef hooks closes the deal, and in a blink the current no longer jams. It makes nice, joining those of us who ride as Neptune intended.

3 minutes at 15' to ditch excess nitrogen comes next, but at 4 knots the bottom turns deep blue in a blink, and I hear Tutor the Turtle in the offing: *Help, Mr. Wizard!*

Calmness. Faith. Neptune loves you.

Relaxed, trimmed out to neutral buoyancy, it's time for some casual portraits of my current mates, whose composure feels profoundly assuring. Gee, it must be close to lunchtime, I thought. All this current makes a swimmer hungry. Hmm… At least everyone along for the ride is in the same boat, or in the same ocean, which may be more comforting to the big galoot up top waiting to say hello.

Then it's back on board for a rousing round of hail-fellow, and hey, did you see that orange-finned and pink anemonefish putting on the show?

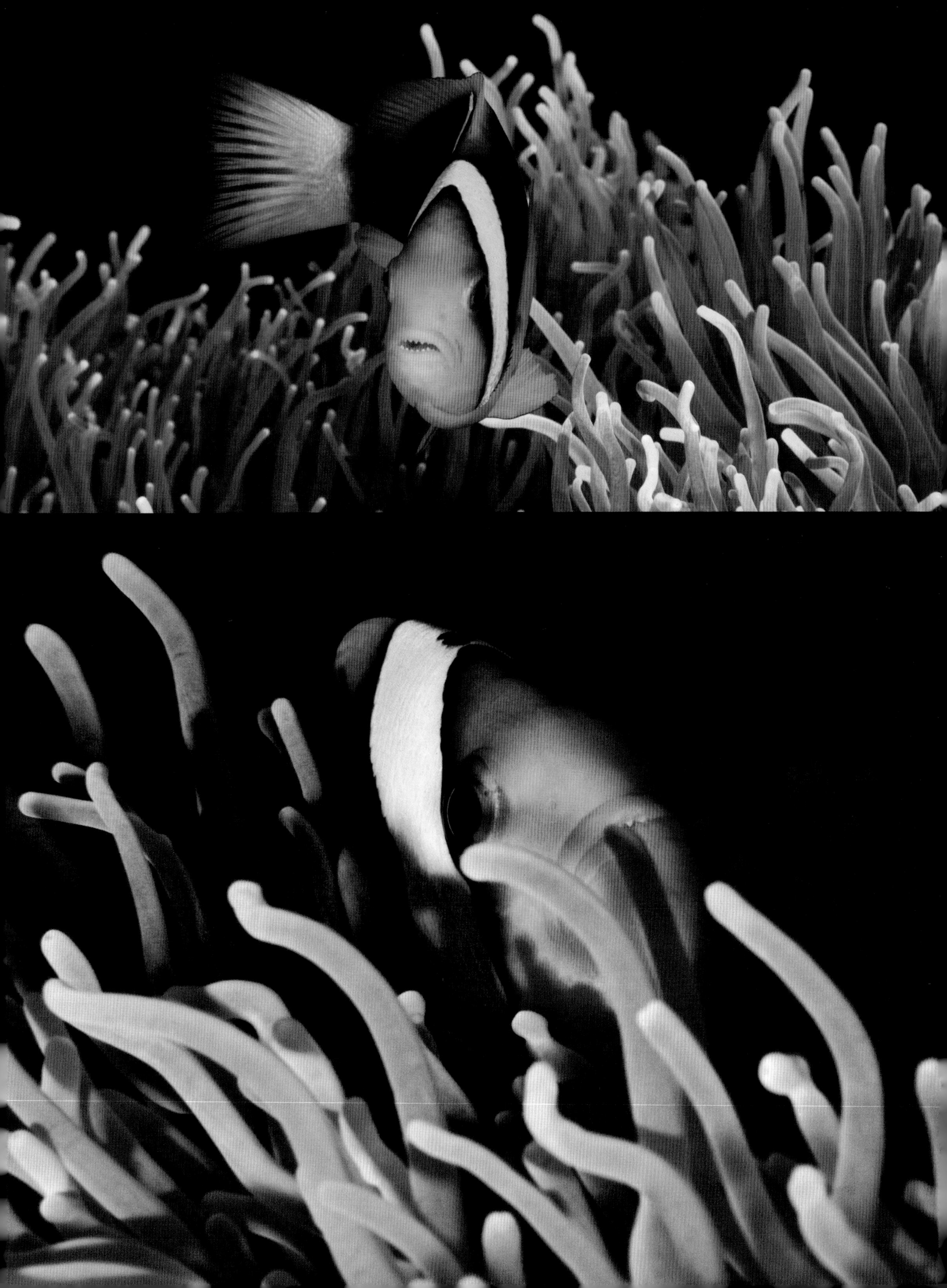

Orange-finned anemonefish (above), pink anemonefish (below)

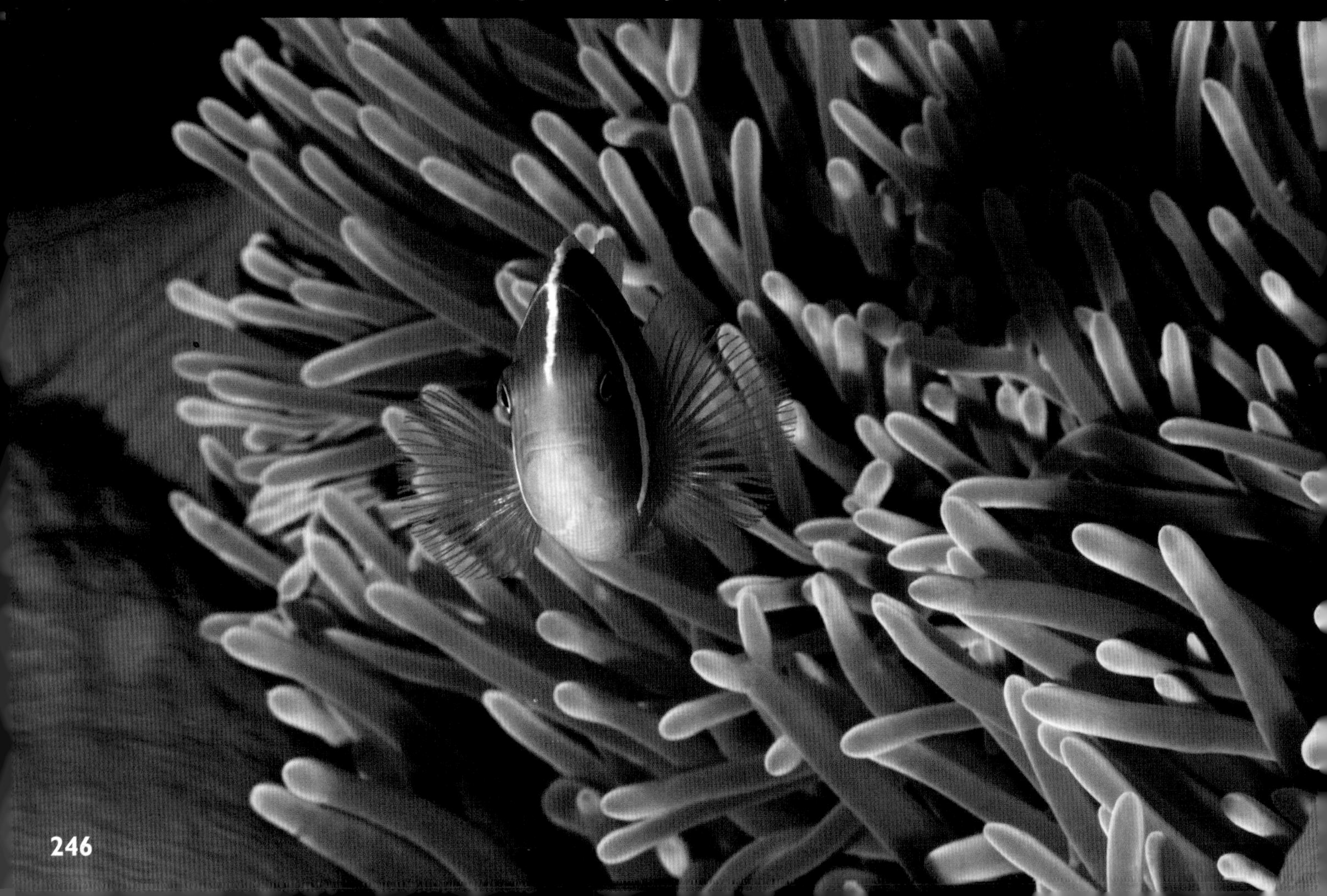

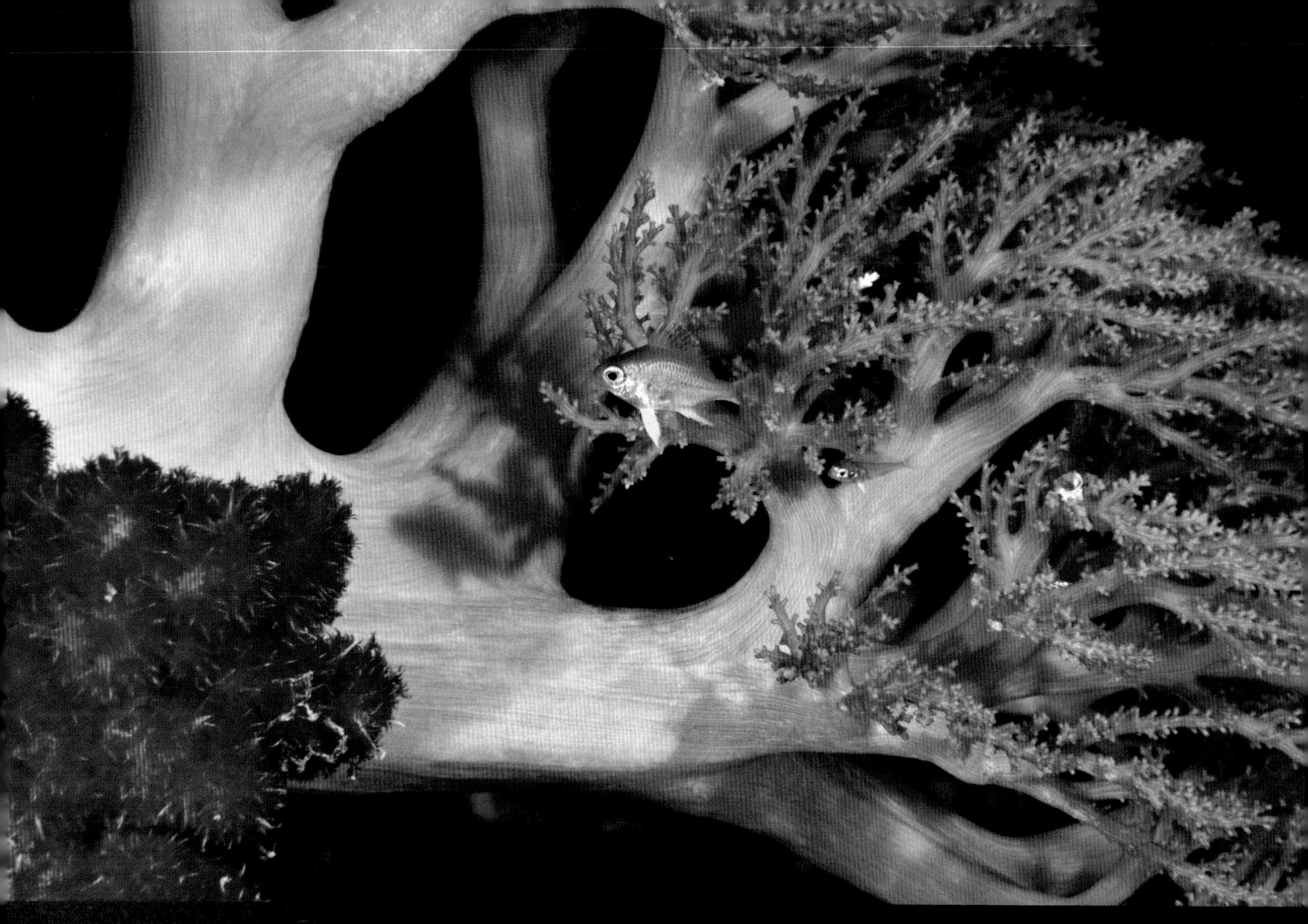

Indelible images of simple truth and the meaning of life may soothe pedestrian lives from a more structured, angular society based on material gain. Can a dull commute ever be the same, once warmed by Neptune's light?

All reverie ends bye'm bye, this one on a tough question posed by a fellow who wants to know: how in hell could the tiny republic of Palau cure itself of the aquarium scourge, when the sovereign free State of Hawaii cannot? Of course the fellow is me, Snorkel Bob, giving voice to a few gill breathers who still wonder…

A 2-spined angelfish (above) wants to know.

A bridled monacle bream (below) wants to know.

A reticulated butterfly (above) and hawk anthius (right) wait for elucidation, and a flagtail grouper (below) claims a vested interest in this foodchain—I mean neighborhood.

A dotted butterflyfish (above), a Bennett's butterflyfish (right), and a Pacific double-saddle butterflyfish (below) await the answer.

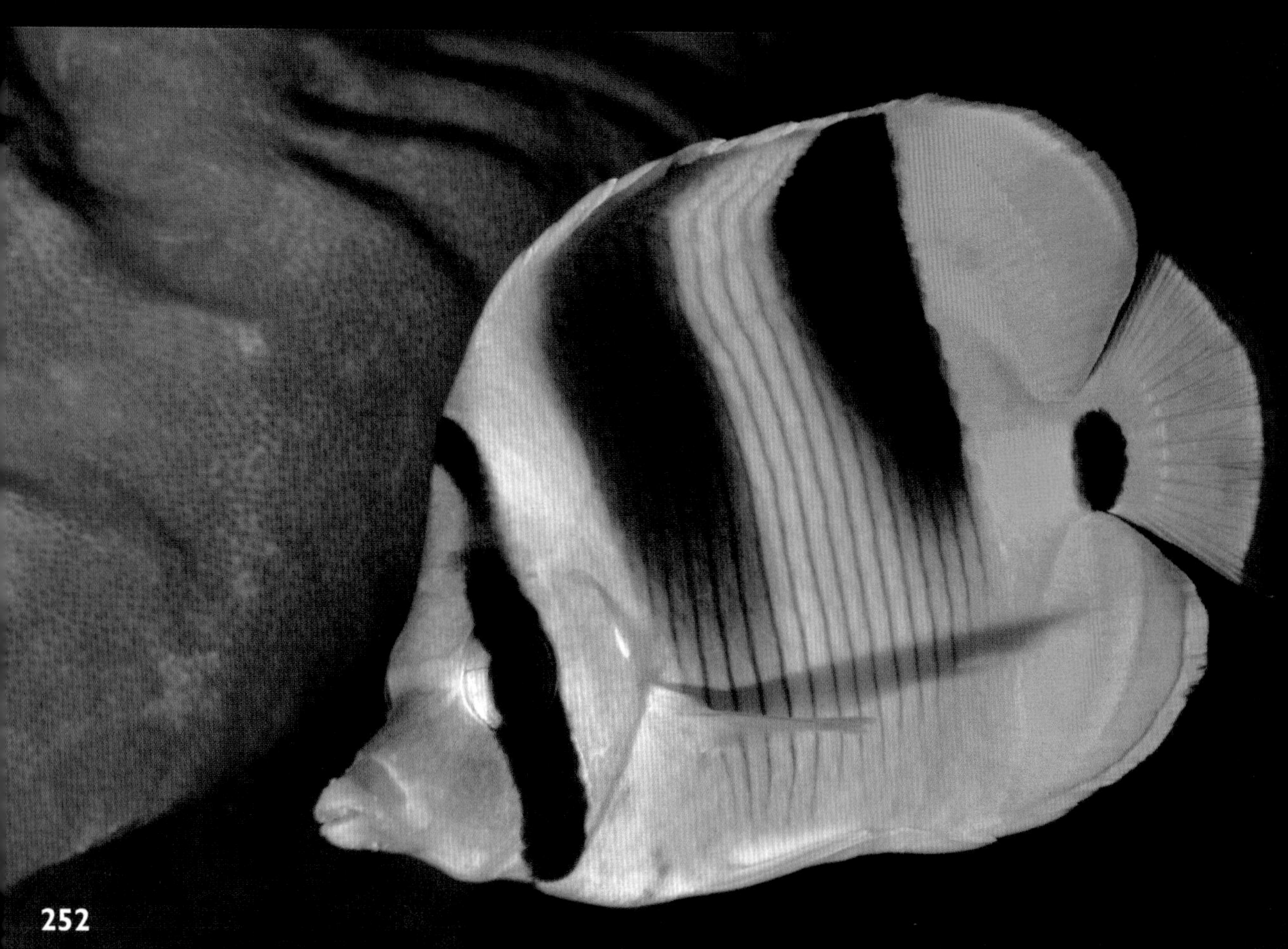

Chromis ultimatus:

We need to know how safe we are and if the aquarium scourge will come again to put things out of whack. We can't spawn on devastated reefs or reefs out of balance, sacrificed to commercial extraction.

Sabre squirrelfish:
We're not pussyfooting around!

Wait! I know what happened. Okay, this guy, a Palau guy, was making money selling us into the aquarium hobby…

…and the state governments across Palau allowed trafficking in wildlife for the pet trade—as long as the traffickers would pay each state a tariff. That's how it goes on Planet $. Things balance naturally, but with human commerce, nature dies for dollars, and we take it in the anus every time.

What does a longfin spadefish know? I heard on the coconut wireless that the Palau guy got help from another guy with the connections to wipe out Palau reefs for top frikkin' dollar for the aquarium trade AND for Asian restaurants demanding smaller reef fish, because they ate all the other fish. Poaching, schmoaching; they don't give a batfish patootie. They'll eat anything, if they think it's a last gasp and the honor is all theirs.

I heard that the issue was an easy choice for the Palau Congress, with only one Palauan making money on the devastating extraction, and another guy from Taiwan or Korea or Timbuktu siphoning moolah like no tomorrow. Yeah, they were sloppy, pulling up scads o' fish in plain view of hundreds of tourist divers still in the glow of reef communion—divers spending about a grand a day each on average just to visit, view and leave, taking nothing away but memories.

The masterstroke came from President Remengesau. The Congress was itching for bigger loans. The bank-loan bill was huge. We looked like a little mud minnow hooked on for the ride. That's how it works with reefs. You get lucky on the timing or you don't.

You're both correct in general terms, Bucky and Sam, but don't discount a good motivation, rare as it may be. A dive operator turned the tide by posting photos of a factory ship servicing a fishing fleet presumably from Taiwan, Korea or Hong Kong, strip-mining Palau reefs, as if we were free for the taking—as if we weren't working our tails off to support the economic engine of this republic.

Call it politics if you must; I can live with that. I call it heroic. President Thomas Remengesau, Jr. (1/01 to 1/09) saw revenue for the states but none for the republic of Palau, and he saw the end of a dive destination regarded around the world as premium. He slipped our rider onto the bill for bigger loans from the Palau National Development Bank. Just that quick, it was over—no floor fight, no cries of woe like you hear in Hawaii. Live fish exports from Palau were, are and shall be illegal.

It all seems terribly melodramatic to me.

I'd like to see how melodramatic you'd feel in some lemon butter caper sauce. Or worse, flopping on a table while they slice you thin for sashimi with a pulse. Don't they know about edamame? President Remengesau called it "the insatiable markets of Asia."

Yes, they're intrigued, disemboweling beautiful creatures. What fries my oysters are the comments you see from the armchair quarterbacks on-line. One guy said that stopping reef genocide will affect the livelihood of the reef killers. Another guy called that point valid, saying conservation must consider the needs of the local population. We're the local population. The two who started this crime made good dough. The rest worked for chump change. Wake up and smell the detritus!

Okay, Sweetlips. Good to go.

Geez, Little Blue, when did you get so tough? And when do you go on-line?

Hey! Soldierboy. Get over here!

Holy Neptune, Soldierboy! Talk about eating anything. This is disgusting…

W*ell we all need…someone…we can…clean on…* Oh, darn, now I won't get that song out of my head. But that's *manini* next to knowing the aquarium scourge AND a voracious restaurant appetite has been blocked from killing Palau reefs—sustainably, of course. So a host of critters wait to entertain on reefs regarded around the world as premium—as protected forevermore or until the next challenge from the single species outracing its headlights.

Just for perspective, parrotfish are a lynchpin species critical to reef survival, yet humans eat them with no constraint in Palau. This is a humphead parrot.

photo by Anita

Also harsh: hawksbill turtles are endangered worldwide, yet Palauans kill them to make tourist chachkas from their shells. Open season is every other quarter, so ½ the hawksbill population will always be intact. What? That same idiotic gobbldegook explained green sea turtle harvesting in Hawaii decades ago. Wake up, Palau! Above, a hawksbill plastron (bottom casing), cut away with a knife and left on the bottom.

Right, a whitecheek surgeon grazes on algae. (In Hawaii it's called a goldtrim surgeon.) Below, a Pacific sailfin tang. Herbivores keep algae in check.

Shrimp gobies, above, carry on the debate. Below, a cleaner wrasse cleans a Meyer's butterfly.

All of life came from the sea, and so did love.

photo by Anita

Above, a leaf scorpionfish with a scenic view. Below, a pink anemonefish youth confides:

A Hui Hou...

from left: Lulu and Robert Wintner

About the Author

Robert Wintner is the Executive Director of the Snorkel Bob Foundation and works diligently to protect Hawaii reefs. He led legislative campaigns at the Hawaii State Capitol and in Maui County from 2007 through 2011. First passing the State Senate with unanimous consent, the aquarium campaign got derailed in the House by vested interests in leadership and the Executive Branch. These efforts brought aquarium plunder into the light of day, however, with a grim reality and empty reefs facing the people of Hawaii. Legislative efforts will continue.

Wintner's short fiction has appeared in Hawaii Review, (U. of Hawaii) and Sports Illustrated. His 2007 novel, *In a Sweet Magnolia Time* was nominated for a Pulitzer Prize. Wintner's adventure novel of the sailing charter trade *Whirlaway* and *Modern Outlaws* were both optioned for film rights by Hollywood production companies. Other novels and accolades over the years led to *Some Fishes I Have Known*, a first glimpse at the social side of the reef community.

Every Fish Tells a Story brings into focus some key personalities from reefs across the western world, along with insight on the devastating aquarium trade. Asked in a radio interview if a fish can have a soul, Wintner said, "I wonder if some people have souls. The question of souls and who might have one is strictly human. The question of souls is a luxury, a mind game, and it's demeaning in the sense that fundamentalists can deny souls to some species. Humans tend to define intelligence in other species as a measure of the ability to learn certain behaviors. But many species decline to learn those behaviors and still have lessons to teach—to those of us willing to learn. So let me rephrase the question: 'Can a fish be a friend of mine?' The answer is yes, obviously. Look at these pages."

Neptune Speaks goes to the heart of the matter—to the primary stakeholders on every reef around the world, giving them a voice on the truth. "I start to feel like Johnny One Note," Wintner said. "Calling foul on the aquarium trade over and over again. But reef plunder is a topic so rife with greed and evil and the very worst behaviors of humanity that the note takes on a life of its own. At a certain age we might learn to stop resisting what comes naturally to us, so I don't worry anymore about other issues I find compelling. The aquarium trade mentality killed the Carolina parakeet, the American buffalo, the Steller sea lion and many more—it's a mentality that wants to save the world, unless killing off just a little bit of it for personal gain would be okay."

Wintner is still a voice for marine mammals, turtles, the reefs and its citizens and vows to continue. "It's a long march. We have no choice but to take the next step."

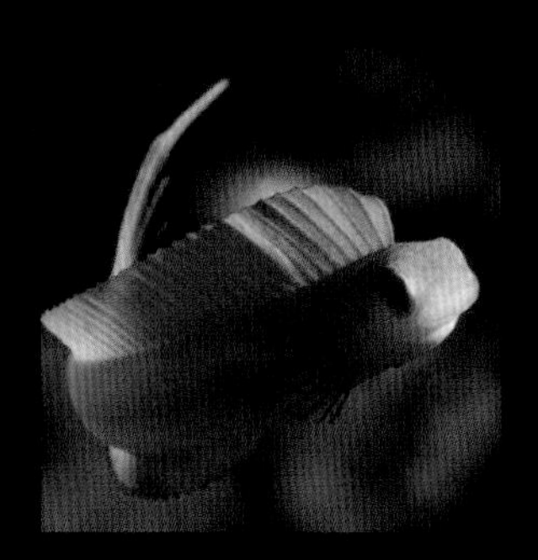

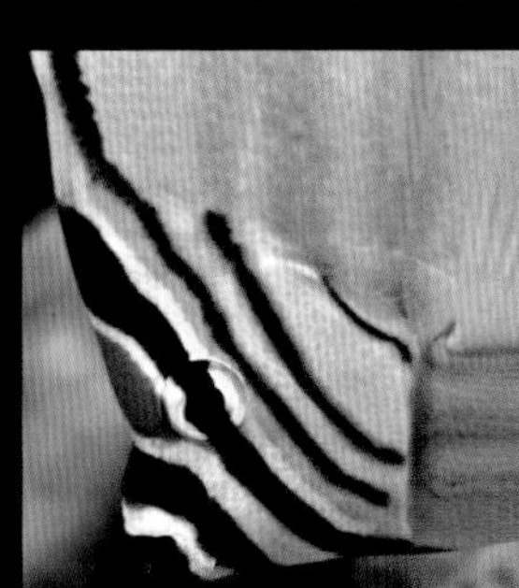

Flame Angels are iconic in the anti-aquarium campaign around the world.

New from Robert Wintner

Flame Angels is a tale of adventure on the high seas. Ravid's 20 years of leading dives in Hawaii have rendered the tanks heavier; the women are still temporary, and the abounding beauty is growing scarce. Enter Minna Somayan with fresh perspective. They marry in two weeks.

Flame Angels is a tiderip of character and setting. A dive leader facing middle age cannot reconcile society and nature, until he must. Consider a waterman miles from shore in the ocean at night. Fear would grip any man who fails to reconcile the unseen presence as a guiding spirit.

Just for fun, turn the page and have a little *pupu*:

Flame Angels

a novel of Oceania

Robert Wintner

Prologue—Halfway Out to Sea

A lean and sun browned man slithers in the shallows easily as an eel after fry, till he draws his legs under and stands, taller than the first organisms walking out of the sea but with original intent: to improve his niche, on land.

With his hair wetted to his neck, a scant loin pocket and a scruffy beard dripping under swim-goggle eyes, he makes amphibious transition, coming up and out.

An emergence from an hour or two of reef repose gives pause for wonder. Ankle deep he watches two children and a dog playing naked in the waves. The girl of ten sweeps her long, black hair out of the way childishly with both hands. Composed as a tropical cameo one moment, she surges with energy the next, yelling at the boy to eat. Mangez! C'est une tempête en mer, et vous devez manger pour rester forts! It's a storm at sea, and you must eat to stay strong!

He sits on a paddleboard. She pulls it by a rope along the shallow surf line.

The dog barks, trying to jump on and finally clambers aboard, where he teeters, facing back to better watch the boy.

The boy eats from a plate on his lap: baked yams, carrots and pineapple. Small wave-tops season his lunch. The dog whines.

Leihua pulls the paddleboard to waist depth and points it into the surf, then gives it a push, commanding him to eat.

Justin eats, piercing the short break, focused on the pineapple saved for last along with a piece of taro for the dog.

The man says, "Mes enfants," as a statement of being, a navigational fix on terra firma. He slogs up a sandy path, no longer buoyant. The kids and dog gravitate and follow toward the house, leaving the paddleboard high on the beach.

At the house they'll rest and pass the hot afternoon. They might doze. In two hours the boy and girl will tend to schoolwork while the man prepares dinner. He may work images on his computer for a while before the woman arrives from the hospital.

Meanwhile, on the way up, they pass a mound of dirt topped with smooth rocks, the topmost a marker, engraved: Skinny. It's a final resting place, but its reluctant tenant would rather use it as a perch. She jumps to the top and wobbles off, so the man picks her up and sets her on top again. The old, feeble cat with the baby face suddenly sees him and speaks her catch-all word to the omniscient one who insists that she keep living, that she keep processing moments as she has the last twenty-two years.

When the man and children pass, she leans forward to swat the dog on the butt but falls off trying and meows again, falling into the procession up the path.

(*Flame Angels*, Iguana Books, Toronto, 2012)

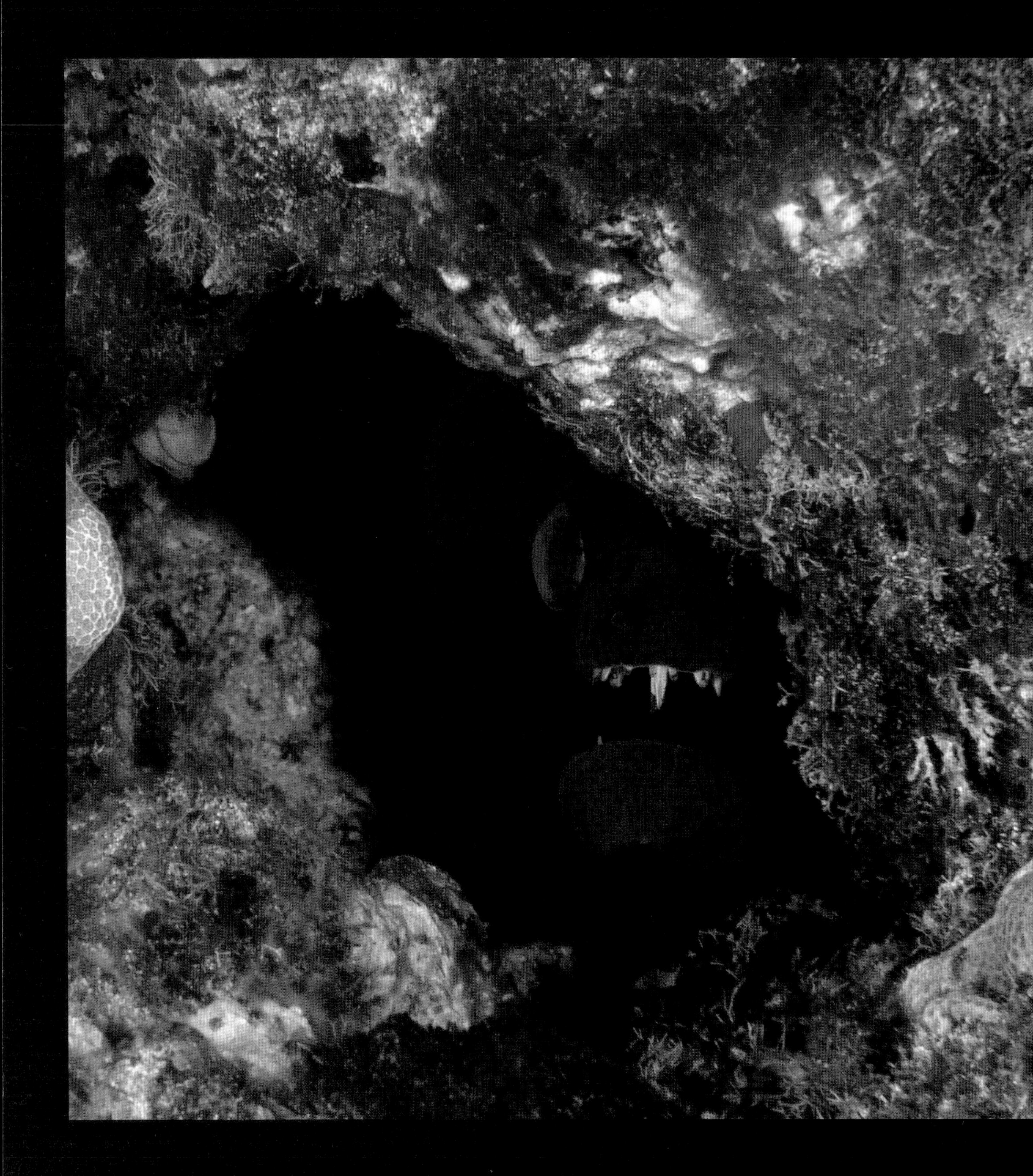

Index

Symbols

A

B

C

D

E

V

W

Y

Z

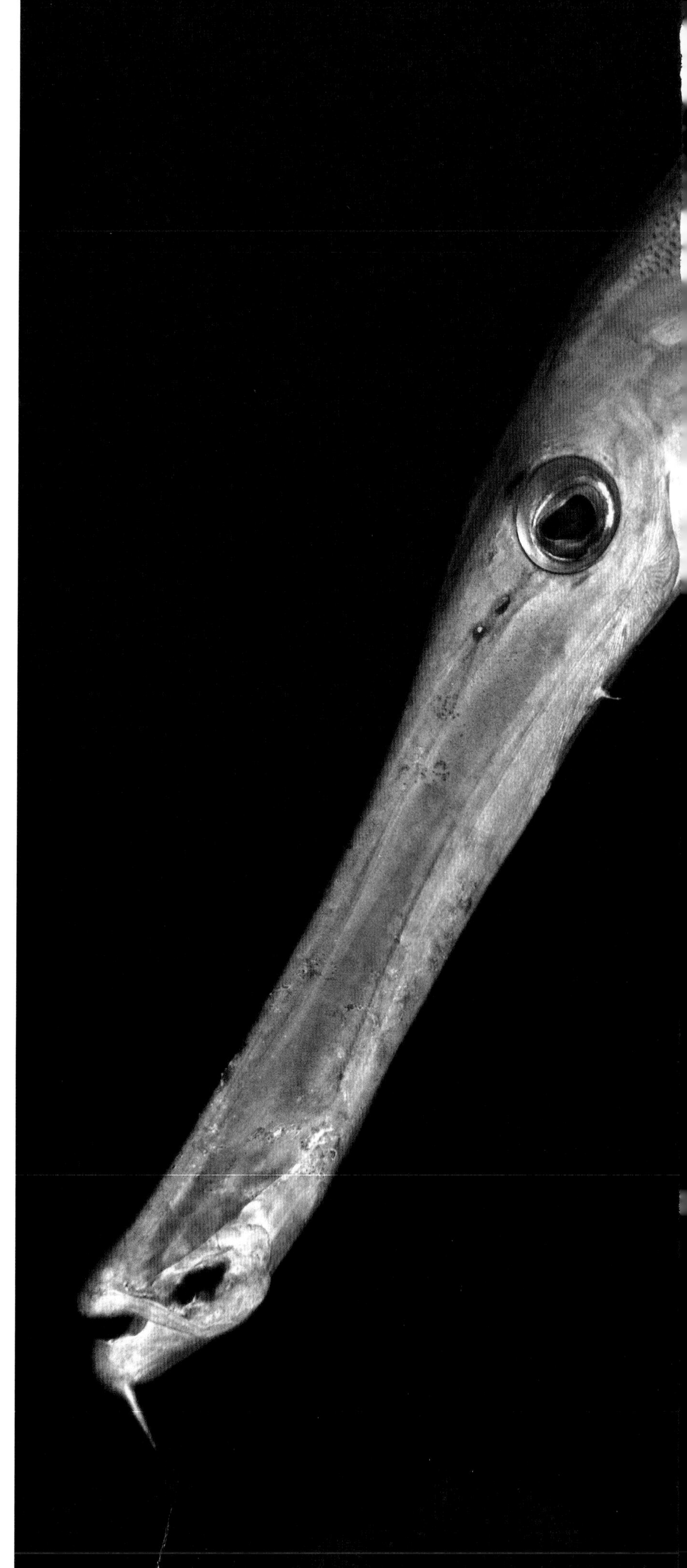